T0316458

Stochastic Analysis of Mixed
Fractional Gaussian Processes

Series Editor
Nikolaos Limnios

Stochastic Analysis of Mixed Fractional Gaussian Processes

Yuliya Mishura
Mounir Zili

First published 2018 in Great Britain and the United States by ISTE Press Ltd and Elsevier Ltd

ISTE Press Ltd
27-37 St George's Road
London SW19 4EU
UK

www.iste.co.uk

Elsevier Ltd
The Boulevard, Langford Lane
Kidlington, Oxford, OX5 1GB
UK

www.elsevier.com

British Library Cataloguing-in-Publication Data
A CIP record for this book is available from the British Library
Library of Congress Cataloging in Publication Data
A catalog record for this book is available from the Library of Congress
ISBN 978-1-78548-245-8

Printed and bound in the UK and US

Contents

Preface

This book reflects upon the current trends that arise in both the theoretical study and the practical modeling of phenomena that exhibit certain properties of fractionality. That is, there is a tendency to transition from one type of fractionality to several types that are present at the same time. In this connection, there is a need to construct models that include different types of fractionality. Various approaches to the construction of such models can be considered; however, in our opinion, one of the simplest approaches, suitable for applications in engineering as well as in economics and finance, is the modeling of complex fractional properties using the so-called mixed models, or, more simply, linear combinations of some elementary fractional processes. In this case, we can adequately model different levels of fractionality as well as not complicate the model too much. On the basis of these arguments, we have devoted this book to the mixed fractional Gaussian processes, that is, the linear combinations of fractional Brownian motions and sub-fractional Brownian motions. The book is a textbook as well as a monograph, because, on the one hand, we tried to make it useful for beginners and included the necessary information from analysis, the theory of random processes, the theory of Gaussian processes, fractional calculus, Hausdorff and capacitarian dimensions, etc.; and on the other hand, we wanted to make it useful for researchers, for whom we presented our latest theoretical achievements in this field.

The book consists of three main chapters and an Appendix. Chapter 1 is devoted to general information from the theory of Gaussian processes. We have taken only the necessary information from this vast theory. In addition to the general definitions, the main issues that are presented here are: a variant of the theorem on normal correlation with necessary conclusions; continuity and Hölder properties of Gaussian processes; uniform bounds and maximal inequalities and, finally, equivalence and singularity of measures related to Gaussian processes. We tried to provide standard facts with detailed comments on the possible properties of the principal characteristics of Gaussian processes, such as incremental distance, properties of covariance matrix, linear independence of the values of Gaussian processes and the corresponding properties of the induced measure, in order to make the study transparent and show the development of the Gaussian property over time. Chapter 2 is devoted to fractional and sub-fractional Brownian motions. For a sub-fractional Brownian motion, existence and the main properties are studied, together with various representations including moving average, spectral and compact interval representations. Also discussed are long- and short-range dependence of fBm and sfBm, Wiener integration with respect to these processes, asymptotic growth of trajectories and the Girsanov theorem for a sub-fractional Brownian motion. Chapter 3 is the most extensive, because it is devoted to the main subject of this book, namely mixed fractional and mixed sub-fractional Brownian motions. Non-degeneracy of such combinations, properties of their trajectories, series expansions, some computer simulations and semi-martingale and non-semimartingale properties are studied. The chapter concludes with a study of the properties of a mild solution of mixed sub-fractional colored-white heat equation. All main chapters are supplied with exercises, implementation of which will help the reader to better understand the proposed theory. The Introduction contains an extended description of the previous research devoted to fractional and sub-fractional Gaussian processes. The Appendix contains the necessary preliminary information from algebra, calculus and stochastics.

To summarize, this book will be useful to specialists in the field of fractional processes, as well as to practitioners, students, graduate students and readers interested in applications of fractional random processes.

Yuliya MISHURA
Mounir ZILI
March 2018

Introduction

Fractional Gaussian processes are considered widely for the following two reasons: (1) the theory of fractional Gaussian processes is quite interesting in itself and (2) it has extensive applications. Fractional Brownian motion (fBm) seems to be the simplest fractional Gaussian process. It was introduced in the pioneer paper [KOL 40], and has since been widely studied and has numerous applications in the fields of economy, finance and engineering. For example, in the opinion of engineers, fBm has been successfully applied in fields of engineering such as characterization of texture in bone radiographs, image processing and segmentation, medical image analysis and network traffic analysis. Its peculiarity is the presence of the so-called memory, that is, the absence of the Markov property, which manifests itself in all fairly complex dynamic systems. For all these reasons, the theory of fBm was considered to the smallest detail. We mention here only the books and some extended papers devoted to fBm and its applications, including [DOU 02, SAM 06, BIA 08, MIS 08, NOU 12, MAN 97], to name a few.

However, fBm alone cannot serve as an adequate model in all spheres of applications, and more complex fractional random processes are needed to model real phenomena. On the one hand, the fractional Brownian motion, which is characterized by a single parameter, namely the Hurst index, cannot serve as a good model where there are several levels of fractionality. Thus, a mixed model can be introduced as a linear combination of different fractional Gaussian processes. The simplest case is a mixed model based on both standard and fractional Brownian motions, which turns out to be more flexible. One of the reasons to consider such a model comes from the modern mathematical finance, where it has become very popular to assume that the underlying random noise consists of two parts, namely the fundamental part, describing the economical background for the stock price, and the trading part, related to the randomness inherent to the stock market. The fundamental part of the noise has a long memory, while the trading part is a white noise. Such mixed fractional Brownian motion was introduced in [CHE 01] to present a stochastic model of the discounted stock price in some arbitrage-free and complete financial markets, and since then it has been sufficiently well studied. The model has the form $M = B + \alpha B^H$, where B is a Brownian motion, B^H is an independent fractional Brownian motion with Hurst index $H \in (0, 1)$ and α is a non-zero real number. It was shown that M is not a semi-martingale if $H \in (0, 1/2) \cup (1/2, 3/4]$, and is equivalent in measure to a Brownian motion if $H \in (3/4, 1)$. From this point on, the mixed fBm has been comprehensively studied. In particular, the absence of arbitrage in the class of self-similar strategies was established in [AND 06]; the existence, uniqueness and properties of the solution of the stochastic differential equation involving Wiener process and fractional Brownian motion with Hurst index $H > 1/2$ were studied in [MIS 11a, MIS 11b, MIS 12, MEL 15]. In [ELN 03], a modification of the mixed fBm, the so-called fractional mixed fractional Brownian motion (FMFBM), was introduced and the lower-lower class of non-random functions (one of the Lévy classes) for such a process was characterized. In [ZIL 06], a slightly different mixed fractional Brownian motion $M = aB + bB^H$, where a, b are real numbers, was considered, and the fractional differentiability of its sample paths was investigated.

Parameter estimation in the mixed Brownian–fractional Brownian model is provided in [ACH 10, CAI 16, DOZ 15, FIL 08, KOZ 15, XIA 11]. These techniques are described in [KUB 17].

It is possible to say that financial and other models also very often demonstrate the phenomena of multifractionality, where the depths of the memory vary with time; however, such multifractional processes will not be the subject of our consideration. We refer the readers interested in multifractional processes to [BEN 98, AYA 00, COE 05].

The goal of this book is, however, slightly different. First, we consider a generalization of Brownian motion, namely sub-fractional Brownian motion. Second, we model multifractionality using linear combinations of fractional and sub-fractional Brownian motions. For the mixed models with different fBms of the form $M = \sum_{i=1}^{N} a_i B^{H_i}$, where B^{H_i} are independent fractional Brownian motions and $a_i \in \mathbb{R}$, see, for example, [THÄ 09, MIA 08] and section 1.6 in [PRA 10] and the references therein.

Sub-fractional Brownian motion is a centered Gaussian process, intermediate between Brownian motion and fractional Brownian motion. It has some of the main properties of fractional Brownian motion such as self-similarity and Hölder paths, and it is neither a Markov process nor a semi-martingale. However, its increments are not stationary; they are more weakly correlated on non-overlapping intervals and their covariance decays polynomially at a higher rate as the distance between the intervals tends to infinity; for this reason, this process is called sub-fractional Brownian motion. Sub-fractional Brownian motion (sfBm) was introduced in

[BOJ 04], in which the existence and some properties of sfBm, such as long-range dependence, self-similarity and non-stationarity, were established. It was demonstrated how sfBm arises from occupation time fluctuations of branching particle systems and the long memory effect of the initial condition was exhibited. Some more stochastic properties of sub-fractional Brownian motion were given in [TUD 07, TUD 11]. More precisely, in the former, Constantin Tudor studied strong variation, re-normalized power variation, Dirichlet property, as well as short or long memory properties of sfBm. In the latter, the author derived explicit bounds for the Kolmogorov distance in the central limit theorem (CLT) and proved the almost sure CLT for the quadratic variation of the sfBm. The above-mentioned properties make the sfBm a possible candidate for models involving long-range dependence, self-similarity and non-stationarity. This process also appeared independently in a different context in [DZH 04]. Because, unless it is not a Brownian motion, the sfBm is not a semi-martingale, many of the power techniques from stochastic analysis are not available when dealing with it. Despite these difficulties, many authors dealt with this subject. Indeed, because sfBm is a Gaussian process, it is possible to construct a stochastic calculus of variations with respect to it (see, for example, [ALÒ 01]). Because sfBm has Hölder paths, in [TUD 07], the author defined the stochastic integral with respect to it in the path-wise sense; studied the asymptotic behavior of the realized power variation of the corresponding stochastic integral; and established the convergence in probability, uniformly on compact time intervals, of some variations of the indefinite integral with respect to sfBm. In [TUD 09], the domain of the Wiener integral w.r.t. a sub-fractional Bm was described separately for $H > 1/2$ and $H < 1/2$.

In [SHE 10], a stochastic calculus for sfBm with $H > 1/2$ was developed by using the techniques of the Malliavin calculus. The bounds in L^p and maximal inequalities for the divergence integral with respect to sfBm were established. Stochastic calculus connected to sub-fractional Brownian motion, also with $H > 1/2$, was considered in [YAN 11], where the authors focused on obtaining various versions of Itô's formula and introduced the integral of deterministic functions with respect to the local time of sfBm. In [SHE 11], the p-variations of some integral functionals connected to the local times of a sub-fractional Brownian motion with index $H \geq 1/2$ were studied. [LIU 12b] was devoted to producing, by means of the Malliavin calculus, the convergence results for some properly re-normalized weighted quadratic variations of sub-fractional Brownian motion with index $H < 1/4$. In the same year, it was proved in [PRA 12] that the probability measures generated by two sub-fractional Brownian motions with different Hurst indices are singular with respect to each other. In [LIU 12a], the authors studied the problem of self-intersection local time of d-dimensional sub-fractional Brownian motion on the basis of the property of chaotic representation and the white noise analysis. In [SHE 14a], an approximation theorem for sub-fractional Brownian motion with $H > 1/2$ was obtained using martingale differences.

Recently, in [PRA 17a], the author obtained a maximal inequality for sub-fractional Brownian motion with Hurst index $H > 1/2$, analogous to the Burkholder–Davis–Gundy inequality for fractional Brownian motion derived in [NOV 99], and established an integral inequality for Wiener integrals with respect to a sub-fractional Brownian motion with Hurst index $H > 1/2$. The existence, the joint continuity and the Hölder regularity of local time of sfBm were established in [MEN 10], where Chung's form of the law of iterated logarithm for sfBm was given. An Ornstein–Uhlenbeck process (OUP) driven by a sub-fractional Brownian motion was considered in [YIN 15]. It was proved that a sub-fractional Ornstein–Uhlenbeck process is locally non-deterministic. As an application, assuming that $d \geq 2$, it was established that the intersection local time of two independent, d-dimensional sub-fractional Ornstein–Uhlenbeck processes with the same Hurt index H exists in L^2 if and only if $Hd < 2$.

Statistical problems with sub-fractional Bm, especially parameter estimation in the models involving sfBm, have been investigated widely. In particular, in [LIU 12c], the authors presented two convergence results about the second-order quadratic variations of sub-fractional Brownian motion: the first is a deterministic asymptotic expansion; the second is a central limit theorem. Then, they combined these results and concentration inequalities to build confidence intervals for the self-similarity parameter associated with one-dimensional sub-fractional Brownian motion. The parameter estimation problem for the sub-fractional OUP, in the case of a Hurst index $H > 1/2$, was considered by Mendy in [MEN 13], where he studied the consistency and the asymptotic distribution of the least-squares estimator of the unknown parameter. In [SHE 14b], the authors dealt, using a technique based on the Girsanov theorem, with the problem of efficient estimation for the drift of sub-fractional Brownian motion. They also constructed a class of biased estimators of James–Stein type, which dominate, under the usual quadratic risk, the natural maximum likelihood estimator.

In [KUA 15a], the authors investigated the L^2-consistency and the strong consistency of the maximum likelihood estimators (MLE) of the mean and variance of sub-fractional Brownian motion with drift at discrete observation. By combining Stein's method with Malliavin calculus, the central limit theorem and the Berry–Esséen bounds for these estimators were obtained. In [KUA 15b], the authors estimated the drift parameter in a simple linear model driven by sub-fractional Brownian motion. They constructed the maximum likelihood estimator (MLE) for the drift parameter by using a random walk approximation of the sub-fractional Brownian motion and studied the asymptotic behavior of the estimator. A sub-fractional bridge, with Hurst index $H > 1/2$ and depending on an unknown parameter α, was introduced in [KUA 16b], where the asymptotic properties of a least-squares estimator (LSE) for α were also investigated. In [PRA 17b], the problem of optimal estimation of the vector parameter θ of the drift term in a sub-fractional Brownian motion was considered. The author obtained the maximum likelihood estimator as well as the Bayesian estimator when the prior distribution is

Gaussian. By using multiple stochastic integrals and the Malliavin calculus, the authors of [LIU 17] analyzed the asymptotic behavior of the adjusted quadratic variation for a sub-fractional Brownian motion and applied their results to construct strongly consistent statistical estimators for the self-similarity index of sub-fractional Brownian motion. The methods of statistical estimation of the self-similarity index of sub-fractional Brownian motion were summarized and generalized in [KUB 17].

Concerning linear combinations of fractional and sub-fractional Brownian motions, the need for their consideration is dictated by applications to the real processes that exactly demonstrate such properties. Particular examples of such phenomena are: the multi-component fractional quantum Hall effect in graphene studied in [DEA 11], where it was mentioned that the number of fractional filling factors can be three or four; anisotropic Gaussian random fields studied by many authors, see, for example, [BIE 09] and [XIA 09]; and, last but not least, short- and long-term dependences in economy and on financial markets, where financial and economic time series are not stationary and, more importantly, are only invariant to scale over consecutive segments. They consist of super-positions of various self-similar and stationary segments, each with its own Hurst index. Such an absence of global self-similarity is a problem, and the variability of scales can be well analyzed by the simple use of a multi-scalable fractional Brownian motion (in other words, mixed fractional Brownian motion). For more information, see, for example, [DOM 11] and the references therein. Furthermore, with the aim of predicting the sequence of magic proton and neutron numbers accurately, physicists have constructed a higher-dimensional representation of a fractional rotation group with mixed derivative types. This construction leads to the linear combination of three fractional processes with different fractionality; see [HER 10].

Mixed sub-fractional Brownian motion was introduced for the first time in [ELN 15] as a linear combination of a Brownian motion and an independent sub-fractional Brownian motion with Hurst index $H \in (0, 1)$. Its main properties were studied and it was shown that this process can be considered as an intermediate process between a sub-fractional Brownian motion and a mixed fractional Brownian motion. Later, in [ZIL 14], a more general form of the mixed sub-fractional Brownian motion was presented as a linear combination of a finite number ($n \geq 2$) of independent sub-fractional Brownian motions $\sum_{i=1}^{n} a_i \xi_i^{H_i}$. Some basic properties of this process were studied, the non-Markovian and non-stationarity characteristics were shown, under which the conditions it is a semi-martingale, and the main features of its sample paths were investigated. It was also shown that by a suitable variation of the parameters $a_i, i = 1, ..., n$, we can control the level of dependence between the increments of the process, according to the values of the Hurst parameters H_i. In other words, mixed sub-fractional Brownian motion could serve to obtain a good model of a certain phenomenon, taking into account not only the sign (as in the case of the sub-fractional Brownian motion) but also the strength of dependence between the increments of this phenomenon. Then, an explicit series expansion of the mixed sub-fractional Brownian motion was presented in [ZIL 13], where the author also studied the rate of convergence of the obtained expansion. He also presented a computer generation of sample paths for msfBm. The msfBm was also used in [ZIL 16], to introduce a new stochastic heat equation with a colored-white fractional noise, which behaves as a Wiener process in the spatial variable and as mixed sub-fractional Brownian motion in time. A necessary and sufficient condition for the existence of its solution was reported. Regularity properties of this equation with respect to the temporal and spatial variables were also analyzed. Some fractal dimensions of the graphs and ranges of the associated sample paths were investigated.

As can be seen from above, fractional processes are studied in detail both in the sense of theory and in terms of applications. In this book, we wanted to give the reader an idea of both pure and mixed fractional Gaussian processes and teach how to use different models in applications. We leave it to the reader to determine the extent to which it was possible.

1

Gaussian Processes

1.1. Some preliminaries

Let us consider a probability space $(\Omega, \mathcal{F}, \mathbb{P})$. We denote $L^2(\Omega, \mathcal{F}, \mathbb{P})$, or simply $L^2(\Omega)$, the space of square-integrable real-valued random variables $X : \Omega \mapsto \mathbb{R}$, i.e. random variables such that $\mathbb{E}(X^2) < \infty$.

The function $\langle \cdot, \cdot \rangle : L^2(\Omega) \times L^2(\Omega) \mapsto \mathbb{R}$ defined by

$$\langle X, Y \rangle = \mathbb{E}(XY) \text{ for } X, Y \in L^2(\Omega)$$

is an inner product with induced norm $\| \cdot \| : X \mapsto \|X\| = \left(\mathbb{E}(X^2)\right)^{1/2}$ (see definition A.9 and exercise 1.2). Moreover, with the metric induced by the norm $\| \cdot \|$ (see definition A.10), $L^2(\Omega)$ is a Hilbert space (see theorem A.1).

Let us now consider a random variable X defined on the probability space $(\Omega, \mathcal{F}, \mathbb{P})$ and define the space

$$L^2(X) = \{u(X) | u : \mathbb{R} \mapsto \mathbb{R} \text{ is a Borel function}, \ \mathbb{E}(u^2(X)) < \infty\} \quad [1.1]$$

with the norm induced by the inner product in $L^2(\Omega)$. Let the symbol $\mathcal{B}(\mathbb{A})$ denote the sigma-algebra of Borel sets on $\mathbb{A}$.

LEMMA 1.1.– *The space $L^2(X)$ is a closed subspace of the Hilbert space $L^2(\Omega)$.*

PROOF.– Let us consider the probability $\mathbb{P}_X$ defined on $(\mathbb{R}, \mathcal{B}(\mathbb{R}))$ by the formula

$$\mathbb{P}_X(A) = \mathbb{P}(X^{-1}(A)) = \mathbb{P}(X \in A) \text{ for any } A \in \mathcal{B}(\mathbb{R}).$$

Then, $(\mathbb{R}, \mathcal{B}(\mathbb{R}), \mathbb{P}_X)$ is a probability space. Now, consider the functional space $\mathbb{K} = L^2(\mathbb{R}, \mathcal{B}(\mathbb{R}), \mathbb{P}_X)$. That is,

$$\begin{aligned}\mathbb{K} &= \left\{f : \mathbb{R} \to \mathbb{R} \mid f \text{ is a Borel function and } \int_{\mathbb{R}} f^2(x)\mathbb{P}_X(dx) < \infty\right\}\\ &= \left\{f : \mathbb{R} \to \mathbb{R} \mid f \text{ is a Borel function and } \mathbb{E}(f^2(X)) < \infty\right\}.\end{aligned}$$

Consider the function

$$\psi : \mathbb{K} \mapsto L^2(X) : \psi(f) = f(X).$$

Obviously, it is a linear map. Moreover,

$$\begin{aligned}\langle f, g\rangle_{\mathbb{K}} &= \int_{\mathbb{R}} f(x)g(x)\mathbb{P}_X(dx) = \mathbb{E}(f(X)g(X)) = \langle f(X), g(X)\rangle_{L^2(\Omega)}\\ &= \mathbb{E}(\psi(f)\psi(g)) = \langle \psi(f), \psi(g)\rangle_{L^2(X)},\end{aligned}$$

for any $f, g \in \mathbb{K}$. Therefore, ψ is a linear isometry. Because $\mathbb{K}$ is a Hilbert space, it is a closed set. Therefore, $L^2(X)$ is the image of the closed space $\mathbb{K}$ under the isometry ψ; thus, using lemma A.9, we find that $L^2(X)$ is a closed subspace of the Hilbert space $L^2(\Omega)$. □

Throughout this book, we shall use the notations $\mathrm{Var}(X)$ for the variance of a random variable $X \in L^2(\Omega)$, $\mathrm{Var}(X) = \mathbb{E}(X - \mathbb{E}(X))^2$, and $\mathrm{Cov}(X, Y)$ for the covariance of $X, Y \in L^2(\Omega)$, $\mathrm{Cov}(X, Y) = \mathbb{E}(X - \mathbb{E}(X))(Y - \mathbb{E}(Y))$.

1.2. Gaussian variables and vectors

From the vast amount of information concerning the Gaussian distribution and Gaussian random variables, we have chosen and cited here those that will be used in future.

1.2.1. *Gaussian variables*

DEFINITION 1.1.– *A random variable X is called Gaussian if its distribution density with respect to the Lebesgue measure is given by*

$$x \longmapsto p_X(x) = \frac{1}{\sigma\sqrt{2\pi}} \exp\left(-\frac{(x-m)^2}{2\sigma^2}\right),$$

where m and σ denote two constants such that $m \in \mathbb{R}$ and $\sigma > 0$. In this case, we use the notation $X \sim \mathcal{N}(m, \sigma^2)$. In the particular case where $m = 0$ and $\sigma = 1$, X is called a standard Gaussian random variable and is written as $X \sim \mathcal{N}(0, 1)$.

LEMMA 1.2.– *If $X \sim \mathcal{N}(m, \sigma^2)$, then $\mathbb{E}(X) = m$, $\mathrm{Var}(X) = \sigma^2$, and the characteristic function of X is given by*

$$\varphi_X : t \longmapsto \mathbb{E}(\exp(itX)) = \exp\Big(iat - \frac{\sigma^2 t^2}{2}\Big).$$

Moreover, X has the moments of any positive order and they are the functions of m and σ (see exercise 1.3).

LEMMA 1.3.– *Let $X \sim \mathcal{N}(m, \sigma^2)$ and $(\alpha, \beta) \in \mathbb{R} \setminus \{0\} \times \mathbb{R}$. Then, $\alpha X + \beta$ is also a Gaussian random variable, and $\alpha X + \beta \sim \mathcal{N}(\alpha m + \beta, \alpha^2 \sigma^2)$.*

LEMMA 1.4.– *If $X \sim \mathcal{N}(0, \sigma^2)$, then for any $0 < \gamma < 1$, the following negative moment exists:*

$$\mathbb{E}(|X|^{-\gamma}) = C(\gamma)\sigma^{-\gamma} = C(\gamma)(\mathbb{E}(X^2))^{-\gamma/2},$$

where $C(\gamma) = \dfrac{2^{1-\gamma/2}}{\sqrt{\pi}} \displaystyle\int_0^{+\infty} u^{-\gamma} e^{-u^2} du.$

PROOF.– We have $\mathbb{E}(|X|^{-\gamma}) = \dfrac{2}{\sigma\sqrt{2\pi}} \displaystyle\int_0^{+\infty} x^{-\gamma} \exp\Big(-\frac{x^2}{2\sigma^2}\Big) dx$. Thus, by changing the variables $u = \dfrac{x}{\sqrt{2}\sigma}$, we get the proof. □

REMARK 1.1.– Obviously, any moment of negative order, $\mathbb{E}(|X|^{-\gamma}), \gamma \geq 1$, is infinite.

1.2.2. *Gaussian vectors*

Let $(x, y) = x_1 y_1 + \cdots + x_n y_n$, $x, y \in \mathbb{R}^n$ denote the Euclidean inner product in $\mathbb{R}^n$.

DEFINITION 1.2.– *A random vector $X = (X_1, X_2, \ldots, X_n)$ is called Gaussian if for any vector $\lambda = (\lambda_1, \lambda_2, \ldots, \lambda_n) \in \mathbb{R}^n$ random variable $(\lambda, X) = \lambda_1 X_1 + \lambda_2 X_2 + \cdots + \lambda_n X_n$ is Gaussian. In this case, random variables $X_1, X_2, \ldots, X_n$ are called jointly Gaussian.*

If X is a Gaussian vector with expectation (mean) vector

$$M_X = \mathbb{E}(X) = (\mathbb{E}(X_1), \ldots, \mathbb{E}(X_n))$$

and covariance matrix

$$K_X = \mathrm{Cov}(X) = \Big(\mathrm{Cov}(X_i, X_j)\Big)_{1\le i,j\le n},$$

then we use the notation $X \sim \mathcal{N}(M_X, K_X)$. In the particular case where $M_X = 0$ and $K_X = I_n$, an identity matrix, X is called a standard Gaussian vector.

LEMMA 1.5.– *If $X \sim \mathcal{N}(M_X, K_X)$, then its characteristic function is given by*

$$\begin{aligned}\varphi_X(\lambda) &= \mathbb{E}\exp\left(i\Sigma_{k=1}^n \lambda_k X_k\right) = \exp\Big(i(\lambda, M_X) - \frac{1}{2}(\lambda, K_X\lambda)\Big)\\ &= \exp\left(i\lambda^T M_X - \frac{1}{2}\lambda^T K_X \lambda\right) \qquad [1.2]\end{aligned}$$

for any $\lambda \in \mathbb{R}^n$, where λ^T is the transpose vector of λ, if to consider column vectors.

DEFINITION 1.3.– *Two random variables are called uncorrelated if they have the second moments and their covariance is equal to zero.*

LEMMA 1.6.– *Two jointly Gaussian random variables X_1 and X_2 are independent if and only if they are uncorrelated.*

PROOF.– If X_1 and X_2 are independent (not obligatory Gaussian), then

$$\mathrm{Cov}(X_1, X_2) = \mathbb{E}(X_1 - \mathbb{E}(X_1))\mathbb{E}(X_2 - \mathbb{E}(X_2)) = 0.$$

Conversely, let vector $X = (X_1, X_2)$ be Gaussian and X_1, X_2 be uncorrelated. Then, for any vector $\lambda = (\lambda_1, \lambda_2) \in \mathbb{R}^2$,

$$\begin{aligned}\varphi_X(\lambda) &= \mathbb{E}\exp\left(i\Sigma_{k=1}^2 \lambda_k X_k\right) = \exp\Big(i(\lambda_1\mathbb{E}(X_1) + \lambda_2\mathbb{E}(X_2))\\ &\quad -\frac{1}{2}(\lambda_1^2\mathrm{Var}(X_1) + \lambda_2^2\mathrm{Var}(X_2))\Big) = \exp\left(i\lambda_1\mathbb{E}(X_1) - \frac{1}{2}\lambda_1^2\mathrm{Var}(X_1)\right)\\ &\quad \times \exp\Big(i\lambda_2\mathbb{E}(X_2) - \frac{1}{2}\lambda_2^2\mathrm{Var}(X_2)\Big) = \mathbb{E}\exp(i\lambda_1X_1)\mathbb{E}\exp(i\lambda_2X_2), \qquad [1.3]\end{aligned}$$

so X_1 and X_2 are independent.

REMARK 1.2.– Lemma 1.6 and equality [1.3] can be extended to any finite number of Gaussian random variables. Namely, Gaussian vector consists of independent components if and only if the components are uncorrelated. □

LEMMA 1.7.– *If X and Y are two independent random Gaussian variables such that $X \sim \mathcal{N}(m_1, \sigma_1^2)$ and $Y \sim \mathcal{N}(m_2, \sigma_2^2)$, then $X + Y$ is also a Gaussian random variable, and $X + Y \sim \mathcal{N}(m_1 + m_2, \sigma_1^2 + \sigma_2^2)$.*

REMARK 1.3.– Let $X = (X_1, X_2, \dots, X_n)$ be a Gaussian vector. Then, there exist two possibilities:

i) covariance matrix K_X is non-degenerate, i.e. its determinant is a non-zero positive number, and so the inverse matrix exists. This means that $X_1, X_2, \dots, X_n$ are linearly independent, i.e. their linear combination $\lambda_1 X_1 + \lambda_2 X_2 + \dots + \lambda_n X_n$ equals zero if and only if the vector λ equals zero. In this case, the distribution of X is concentrated on $\mathbb{R}^n$, and the joint density of $X_1, X_2, \dots, X_n$ equals

$$x \longmapsto p_X(x)$$
$$= \frac{1}{(2\pi)^{n/2}(\det(K_X))^{1/2}} \exp\left(-1/2(x - M_X)^T K_X^{-1}(x - M_X)\right), x \in \mathbb{R}^n,$$

where $\det K_X$ denotes the determinant of the matrix K_X, and K_X^{-1} is its inverse.

ii) covariance matrix K_X is degenerate, i.e. its determinant equals zero, and so the inverse matrix does not exist. This means that $X_1, X_2, \dots, X_n$ are linearly dependent, i.e. their linear combination $\lambda_1 X_1 + \lambda_2 X_2 + \dots + \lambda_n X_n$ equals zero for some non-zero vector λ. Let the covariance matrix have rank $1 \le k < n$. This means that k random variables from $X_1, X_2, \dots, X_n$ are linearly independent and others are their linear combinations. In this case, the distribution of X is concentrated on $\mathbb{R}^k$.

It follows from remark 1.3 that for any Gaussian vector $X = (X_1, X_2, \dots, X_n)$, we have the following alternative: either there exists $\lambda \in \mathbb{R}^n$ such that $\mathbb{R}^n \setminus \{(0, \dots, 0)\}$

$$\mathbb{P}(\lambda_1 X_1 + \lambda_2 X_2 + \dots + \lambda_n X_n = 0) = 1,$$

or for any $\lambda \in \mathbb{R}^n$, $\mathbb{R}^n \setminus \{(0, \dots, 0)\}$

$$\mathbb{P}(\lambda_1 X_1 + \lambda_2 X_2 + \dots + \lambda_n X_n = 0) = 0.$$

LEMMA 1.8.– *Let vector $X = (X_1, X_2, \dots, X_n)$ be Gaussian with expectation vector M_X and covariance matrix K_X. Also, let C be a non-random $n \times n$ matrix. Then, the vector CX is Gaussian with $M_{CX} = CM_X$ and covariance matrix $K_{CX} = C^T K_X C$.*

PROOF.– Denote c_{ik} elements of matrix C. For any non-random vector $\lambda \in \mathbb{R}^n$ $(\lambda, CX) = (C^T\lambda, X) = \sum_{j=1}^n \widetilde{c}_j X_j$, where

$$\widetilde{c}_j = \Sigma_{k=1}^n c_{jk}^T \lambda_k = \Sigma_{k=1}^n c_{kj} \lambda_k.$$

Comparing this equality with definition 1.2, we get that vector CX is Gaussian. Furthermore,

$$\mathbb{E}((CX)_i) = \mathbb{E}\left(\sum_{k=1}^{n} c_{ik} X_k\right) = \sum_{k=1}^{n} c_{ik} \mathbb{E}(X_k) = CM_X,$$

and

$$\begin{aligned}\mathrm{Cov}((CX)_i, (CX)_k) &= \mathbb{E}\left(\left(\sum_{l=1}^{n} c_{il}(X_l - \mathbb{E}(X_l))\right)\left(\sum_{r=1}^{n} c_{kr}(X_r - \mathbb{E}(X_r))\right)\right) \\ &= \sum_{l,r=1}^{n} c_{il} c_{kr} \mathrm{Cov}(X_l, X_r) = (C^T K_X C)_{ik},\end{aligned}$$

and we get the proof. □

PROPOSITION 1.1.– *Let (X, Y) be a Gaussian vector consisting of two random variables X and Y.*

i) Conditional expectation $\mathbb{E}(Y|X)$ equals

$$\mathbb{E}(Y|X) = a + bX, \text{ where } a = \mathbb{E}(Y) - \frac{\mathrm{Cov}(X,Y)}{\mathrm{Var}(X)}\mathbb{E}(X), b = \frac{\mathrm{Cov}(X,Y)}{\mathrm{Var}(X)}. \quad [1.4]$$

ii) The random variables $Y - \mathbb{E}(Y|X)$ and X are independent.

PROOF.–

i) Denote $\widetilde{X} = X - \mathbb{E}(X), \widetilde{Y} = Y - \mathbb{E}(Y)$. Let us try to define $\alpha_0 \in \mathbb{R}$ such that $\widetilde{Y} - \alpha_0 \widetilde{X}$ and $\widetilde{X}$ become uncorrelated, i.e. $\mathbb{E}(\widetilde{Y} - \alpha_0 \widetilde{X})\widetilde{X} = 0$. Obviously,

$$\alpha_0 = \frac{\mathbb{E}(\widetilde{X}\widetilde{Y})}{\mathbb{E}(\widetilde{X})^2} = \frac{\mathrm{Cov}(X,Y)}{\mathrm{Var}(X)}.$$

Furthermore, Gaussian random variables are uncorrelated if and only if they are independent, see lemma 1.6; therefore, $\widetilde{Y} - \alpha_0 \widetilde{X}$ and $\widetilde{X}$ are independent. Then, it follows from the properties of conditional expectations that

$$\mathbb{E}(\widetilde{Y} - \alpha_0 \widetilde{X}|X) = \mathbb{E}(\widetilde{Y} - \alpha_0 \widetilde{X}|\widetilde{X}) = \mathbb{E}(\widetilde{Y} - \alpha_0 \widetilde{X}) = 0.$$

Therefore,

$$\begin{aligned}\mathbb{E}(Y|X) &= \mathbb{E}(\widetilde{Y} - \alpha_0 \widetilde{X}|X) + \mathbb{E}(Y) + \alpha_0 \widetilde{X} \\ &= \mathbb{E}(Y) - \frac{\mathrm{Cov}(X,Y)}{\mathrm{Var}(X)}\mathbb{E}(X) + \frac{\mathrm{Cov}(X,Y)}{\mathrm{Var}(X)} X.\end{aligned}$$

ii) It is sufficient to prove that $Y - \mathbb{E}(Y|X)$ and X are uncorrelated. But, according to item (i),

$$\mathbb{E}(Y - \mathbb{E}(Y|X))X = \mathbb{E}(YX) - \mathbb{E}\Big(\mathbb{E}(Y) - \frac{\mathrm{Cov}(X,Y)}{\mathrm{Var}(X)}\mathbb{E}(X)$$
$$+ \frac{\mathrm{Cov}(X,Y)}{\mathrm{Var}(X)}X\Big)X = \mathbb{E}(YX) - \mathbb{E}(X)\mathbb{E}(Y)$$
$$- \frac{\mathrm{Cov}(X,Y)}{\mathrm{Var}(X)}\left(\mathbb{E}(X)^2 - (\mathbb{E}(X))^2\right) = \mathrm{Cov}(X,Y) - \mathrm{Cov}(X,Y) = 0,$$

and the lemma is proved. □

COROLLARY 1.1.– *The conditional expectation* $\mathbb{E}(Y|X)$ *is the orthogonal projection of* Y *onto* $L^2(X)$, *where* $L^2(X)$ *is defined by equation [1.1]. That is,* $\mathbb{E}(Y|X)$ *is the unique random variable such that*

$$\mathbb{E}(Y|X) \in L^2(X) \;\; \textit{and} \;\; Y - \mathbb{E}(Y|X) \perp L^2(X),$$

where the symbol $\perp$ *means that* $\mathbb{E}(Y - \mathbb{E}(Y|X))Z = 0$ *for any random variable* $Z \in L^2(X)$. *In particular,*

$$\mathbb{E}\big(Y - \mathbb{E}(Y|X)\big)^2 = \inf_{U \in L^2(X)} \mathbb{E}\big(Y - U\big)^2.$$

REMARK 1.4.– For the general notion of orthogonality in the Hilbert space, see definition A.13.

PROOF.– It follows from lemma 1.1 that $L^2(X)$ is a closed subspace of the Hilbert space $L^2(\Omega)$. Thus, it follows from theorem A.2 that there exists a unique $Z \in L^2(X)$ such that

$$\mathbb{E}(Y - Z)^2 = \inf_{U \in L^2(X)} \mathbb{E}(Y - U)^2 \;\text{ and }\; Y - Z \perp L^2(X).$$

According to [1.4], we have that

$$\mathbb{E}(Y|X) = \mathbb{E}(Y) + \frac{\mathrm{Cov}(X,Y)}{\mathrm{Var}(X)}(X - \mathbb{E}(X)) = aX + b,$$

where $a = \dfrac{\mathrm{Cov}(X,Y)}{\mathrm{Var}(X)}$ and $b = \mathbb{E}(Y) - \dfrac{\mathrm{Cov}(X,Y)}{\mathrm{Var}(X)}\mathbb{E}(X)$. Moreover, X is a Gaussian random variable; therefore, $X \in L^2(\Omega)$. Thus, on the one hand, $\mathbb{E}(Y|X) \in L^2(X)$, and on the other hand, it follows from proposition 1.1 that

$Y - \mathbb{E}(Y|X) \perp L^2(X)$. Therefore, from the uniqueness of the orthogonal projection, we conclude that $Z = \mathbb{E}(Y|X)$ a.s. and we get the stated result. □

COROLLARY 1.2.– *Consider a Gaussian vector* (X, Y). *If we denote* $\mathrm{Var}(Y|X)$ *the conditional variance of* Y *with respect to* X, *that is,*

$$\mathrm{Var}(Y|X) = \mathbb{E}\left(\left(Y - \mathbb{E}(Y|X)\right)^2 \Big| X\right),$$

then $\mathrm{Var}(Y|X) = \min\limits_{U \in L^2(X)} \mathbb{E}\big(Y - U\big)^2$.

PROOF.– It follows from proposition 1.1 that

$$\mathrm{Var}(Y|X) = \mathbb{E}\Big(Y - \mathbb{E}(Y|X)\Big)^2. \qquad [1.5]$$

Combining this with corollary 1.1, we get our statement. □

REMARK 1.5.– We can write equivalently that

$$\mathrm{Var}(Y|X) = \min_{u \in \mathcal{K}} \mathbb{E}\big(Y - u(X)\big)^2,$$

where

$$\mathcal{K} = \{u : \mathbb{R} \to \mathbb{R} \text{ is a Borel function and } \mathbb{E}(u^2(X)) < \infty\}.$$

COROLLARY 1.3.– *For any centered two-dimensional Gaussian vector* (X, Y), *we have*

$$\mathrm{Var}(Y|X) = \inf_{b \in \mathbb{R}} \mathbb{E}\big(Y - bX\big)^2 = \mathbb{E}Y^2 - \frac{\mathrm{Cov}^2(X, Y)}{\mathbb{E}X^2}.$$

PROOF.– On the one hand, using corollary 1.1, we can establish that the random variable $\mathbb{E}(Y|X)$ is the orthogonal projection of Y onto the space $L^2(X)$. On the other hand, because the Gaussian vector (X, Y) is centered, from exercise 3.1, we get

$$\mathbb{E}(Y|X) = \frac{\mathrm{Cov}(X, Y)}{\mathrm{Var}(X)} X.$$

In other words, $\mathbb{E}(Y|X) \in \{bX; b \in \mathbb{R}\} = \mathcal{G}$.

The set $\mathcal{G}$ is clearly a closed subspace of $L^2(\Omega)$. If we denote Y_X the orthogonal projection of Y onto $\mathcal{G}$, we get $\mathbb{E}(Y|X) = Y_X$ a.s. because $\mathcal{G} \subset L^2(X)$, and the orthogonal projection is unique. Consequently,

$$\inf_{b \in \mathbb{R}} \mathbb{E}\big(Y - bX\big)^2 = \mathbb{E}(Y - \mathbb{E}(Y|X))^2.$$

Now, our statement follows from equation [1.5] and also from the simple observation that $\mathbb{E}\big(Y - bX\big)^2$ is a quadratic function of b. □

We will now extend proposition 1.1 to the case of a Gaussian vector

$$X = (X_1, \ldots, X_n)$$

with an arbitrary integer number of coordinates $n \geq 2$.

PROPOSITION 1.2.– *Consider two integers n and p such that $1 \leq p < n$. Let $(X_1, \ldots, X_p, X_{p+1}, \ldots, X_n)$ be a Gaussian vector, Y and Z be two subvectors of X defined by $Y = (X_1, \ldots, X_p), Z = (X_{p+1}, \ldots, X_n)$. If the covariance matrix of vector Z is invertible (or equivalently, non-degenerate), then the conditional expectation $\mathbb{E}(Y|Z)$ equals*

$$\mathbb{E}(Y|Z) = \mathbb{E}(Y) + \Gamma_{Y,Z}\Gamma_Z^{-1}(Z - \mathbb{E}(Z)),$$

where $\mathbb{E}(Y)$ and $\mathbb{E}(Z)$ are the expectation vectors of Y and Z, respectively, and Γ_Z is the covariance matrix of the vector Z:

$$\Gamma_Z = \Big(\mathrm{Cov}(Z_i, Z_j)\Big)_{1 \leq i,j \leq n-p},$$

and

$$\Gamma_{Y,Z} = \Big(\mathrm{Cov}(Y_i, Z_j)\Big)_{1 \leq i \leq p,\ 1 \leq j \leq n-p}.$$

PROOF.– Let us first find a matrix C consisting of real-valued items, with p rows and $n - p$ columns, such that the random vectors $L := Y - CZ$ and Z are uncorrelated. For this, we should have

$$\begin{aligned}\Gamma_{L,Z} &= \Big(\mathrm{Cov}(L_i, Z_j)\Big)_{1 \leq i \leq p,\ 1 \leq j \leq n-p} \\ &= \Big(\mathrm{Cov}(Y_i - (CZ)_i, Z_j)\Big)_{1 \leq i \leq p,\ 1 \leq j \leq n-p} = \Gamma_{Y,Z} - C\Gamma_Z = 0.\end{aligned}$$

Therefore, with $C = \Gamma_{Y,Z}\Gamma_Z^{-1}$, the random vectors $L := Y - CZ$ and Z are uncorrelated. Being additionally Gaussian, they are independent. Therefore,

$$\begin{aligned}\mathbb{E}(Y|Z) &= \mathbb{E}(L + CZ|Z) = \mathbb{E}(L) + CZ \\ &= \mathbb{E}(Y) + C(Z - \mathbb{E}(Z)) = \mathbb{E}(Y) + \Gamma_{Y,Z}\Gamma_Z^{-1}(Z - \mathbb{E}(Z)).\end{aligned}$$

Thus, the proposition is proved. □

As an immediate consequence of proposition 1.2, we get the following statement.

COROLLARY 1.4.– *For* $n \geq 1$, *consider a centered Gaussian vector* $(X_1, \ldots, X_n)$ *such that the covariance matrix*

$$\Gamma_{(X_1,\ldots,X_{n-1})} = \Big(\mathrm{Cov}(X_i, X_j)\Big)_{1\leq i\leq n-1,\ 1\leq i\leq n-1},$$

of the subvector $(X_1, \ldots, X_{n-1})$ *is invertible. Then,*

$$\mathbb{E}(X_n | X_1, \ldots, X_{n-1}) = \sum_{k=1}^{n-1} b_k X_k$$

where the vector $b = (b_1, \ldots, b_{n-1})^T$ *equals*

$$b = \Gamma^{-1}_{(X_1,\ldots,X_{n-1})} c_n,$$

and, in turn, vector c_n *equals* $\Big(\mathrm{Cov}(X_n, X_1), \ldots, \mathrm{Cov}(X_n, X_{n-1})\Big)^T$.

1.3. Gaussian processes

1.3.1. *Some preliminaries concerning stochastic processes*

DEFINITION 1.4.– *Consider a non-empty set* $\mathbb{T}$, *a probability space* $(\Omega, \mathcal{F}, \mathbb{P})$ *and a function of two variables* $X : \mathbb{T} \times \Omega \mapsto \mathbb{R}^d$, *where* $d \geq 1$. *If* $X(t, \cdot)$ *is an* $\mathbb{R}^d$*-valued random variable on* $(\Omega, \mathcal{F}, \mathbb{P})$ *for any* $t \in \mathbb{T}$, *then* X *is called a stochastic process with values in* $\mathbb{R}^d$. *The stochastic process* X *is usually denoted by*

$$X = \{X_t, t \in \mathbb{T}\} = \{X_t(\omega) = X(t, \omega), \omega \in \Omega, t \in \mathbb{T}\}.$$

In the case where $d = 1$, *the process is called real-valued.*

REMARK 1.6.–

i) For fixed $\omega \in \Omega$, the function $X(\cdot, \omega)$ is called a trajectory or a sample path of the process X.

ii) Let $X_t \in L^1(\Omega, \mathcal{F}, \mathbb{P})$ for any $t \in \mathbb{T}$. The expectation function of X is defined by

$$m : t \in \mathbb{T} \longmapsto m_t = \mathbb{E}(X_t).$$

iii) Let $X_t \in L^2(\Omega, \mathcal{F}, \mathbb{P})$ for any $t \in \mathbb{T}$. The covariance function of X is defined by

$$C : (s, t) \in \mathbb{T}^2 \longmapsto C(s, t) = \mathrm{Cov}(X_s, X_t).$$

DEFINITION 1.5.– *The family of finite dimensional distributions (f.d.d.'s) of a stochastic process* $X = \{X_t, t \in \mathbb{T}\}$ *consists of the distributions of all random vectors* $(X_{t_1}, \dots, X_{t_n})$*, for all* $n \geq 1$ *and all* $(t_1, \dots, t_n) \in \mathbb{T}^n$.

DEFINITION 1.6.– *Consider two stochastic processes* X *and* Y.

i) The processes X *and* Y *are said to be indistinguishable if*

$$\mathbb{P}(\omega \in \Omega | X_t(\omega) = Y_t(\omega), \forall t \in \mathbb{T}) = 1.$$

In other words, X *and* Y *are indistinguishable if they have the same sample paths almost surely.*

ii) The processes X *and* Y *are said to be modifications if, for any* $t \in \mathbb{T}$,

$$\mathbb{P}(\omega \in \Omega | X_t(\omega) = Y_t(\omega)) = 1.$$

iii) The processes X *and* Y *are said to be versions if they have the same finite dimensional distributions.*

DEFINITION 1.7.– *Consider a stochastic process* $X = \{X_t, t \in \mathbb{R}_+\}$.

i) The process X *has stationary increments if, for any* $n \geq 1$*, any* $0 \leq t_1 \leq t_2, \dots, t_n, t_i \in \mathbb{T}$ *and any* $h > 0$*, the random vectors* $(X_{t_n} - X_{t_{n-1}}, \dots, X_{t_2} - X_{t_1})$ *and* $(X_{t_n+h} - X_{t_{n-1}+h}, \dots, X_{t_2+h} - X_{t_1+h})$ *have the same distributions.*

ii) The process X *has independent increments if for any* $n \geq 1$ *and any* $0 \leq t_1 \leq t_2, \dots, t_n, t_i \in \mathbb{T}$*, the random variables* $X_{t_2} - X_{t_1}, \dots, X_{t_n} - X_{t_{n-1}}$ *are independent.*

iii) The process is self-similar with index $H > 0$ *(or simply* H*-self-similar), if for all* $a > 0$ *the processes* $\{X_{at}, t \geq 0\}$ *and* $\{a^H X_t, t \geq 0\}$ *have the same finite dimensional distributions. That is, for any* $n \geq 1$, $t_1, \dots, t_n \in \mathbb{R}_+$ *and any* $a > 0$*, the random vectors* $\left(X_{at_1}, \dots, X_{at_n}\right)$ *and* $\left(a^H X_{t_1}, \dots, a^H X_{t_n}\right)$ *have the same distributions.*

PROPOSITION 1.3.– *If* $X = \{X_t, t \geq 0\}$ *is* H*-self-similar process with stationary increments, then*

i) $X_0 = 0$ *a.s.;*

ii) for fixed t, X_t *and* $t^H X_1$ *have the same distribution.*

PROOF.–

i) For each $a > 0$, it follows from self-similarity that the random variables $X_0 = X_{a.0}$ and $a^H X_0$ have the same distribution. Thus, $\mathbb{E}(X_0^2) = \mathbb{E}(a^{2H} X_0^2)$. This means that $\mathbb{E}(X_0^2) = 0$, and consequently $X_0 = 0$ a.s.

ii) It is a straightforward consequence of the self-similarity. □

1.3.2. *Gaussian processes. Definition and main properties. Existence of a Gaussian process with prescribed covariance function*

Consider a non-empty set $\mathbb{T}$.

DEFINITION 1.8.– *A stochastic process $X = \{X_t, t \in \mathbb{T}\}$ is called Gaussian if, for any $n \in \mathbb{N}$ and $(t_1, t_2, \ldots, t_n) \in \mathbb{T}^n$, the random vector $(X_{t_1}, X_{t_2}, \ldots, X_{t_n})$ is Gaussian.*

In other words, X is a Gaussian process if, for any $n \in \mathbb{N}$, $(t_1, t_2, \ldots, t_n) \in \mathbb{T}^n$ and $(a_1, a_2, \ldots, a_n) \in \mathbb{R}^n$, the random variable $a_1 X_{t_1} + X_{t_2} + \cdots + a_n X_{t_n}$ is Gaussian.

REMARK 1.7.– Analyzing the definition of the Gaussian vector and equation [1.2] presenting the characteristic function of Gaussian vector, we see that the law of any Gaussian vector, and consequently, the law of any Gaussian process, is uniquely determined by the knowledge of its expectation function, which for the process is defined as

$$m : t \longmapsto \mathbb{E}(X_t)$$

and its covariance function, namely,

$$C : (t, s) \longmapsto \mathrm{Cov}(X_t, X_s).$$

Let us consider a real-valued function f, defined on $\mathbb{T} \times \mathbb{T} \subseteq \mathbb{R}^2$.

DEFINITION 1.9.– *The function f is called positive semi-definite if, for all $n \in \mathbb{N}$, $t_1, t_2, \ldots, t_n \in \mathbb{T}$ and $a_1, a_2, \ldots, a_n \in \mathbb{R}$, we have*

$$\sum_{j,k=1}^{n} a_j a_k f(t_j, t_k) \geq 0.$$

The following result is the modification for the Gaussian processes of the general Kolmogorov theorem on the existence of a random process with prescribed finite dimensional distributions. Expectation function of a Gaussian process can be an arbitrary measurable function, and the covariance function can be any positive semi-definite and symmetric function. Therefore, the result has the following form.

THEOREM 1.1.– *Let $\mathbb{T}$ be any non-empty set, m be any function from $\mathbb{T}$ into $\mathbb{R}$ and C be any positive semi-definite function from $\mathbb{T}\times\mathbb{T}$ into $\mathbb{R}$ such that $C(s,t) = C(t,s)$ for all $s,t \in \mathbb{T}$. Then, there exists a Gaussian process $X = \{X_t, t \in \mathbb{T}\}$ with expectation function m and covariance function C.*

The proof of this theorem can be found, for example, in [DUD 02] or in [MIS 17].

EXAMPLE 1.1.– *Consider the case $\mathbb{T} = \mathbb{R}^+$ and let $C(s,t) = s \wedge t$ for any $s,t \in \mathbb{R}^+$. Then, the function C is positive semi-definite. Indeed, we note that for any $s,t \in \mathbb{R}^+$ and $s \vee t \le T$,*

$$s \wedge t = \int_0^T \mathbf{1}_{[0,s](u)} \mathbf{1}_{[0,t](u)} du = \langle \mathbf{1}_{[0,s]}, \mathbf{1}_{[0,t]} \rangle_{L^2([0,T],\lambda_1)},$$

where λ_1 is the Lebesgue measure on $\mathbb{R}$ and $\langle \cdot,\cdot \rangle_{L^2([0,T],\lambda_1)}$ denotes the inner product in $L^2([0,T],\lambda_1)$. Therefore, for any $n \in \mathbb{N}$, $a_1,\dots,a_n \in \mathbb{R}$, $0 \le t_1 \le \dots \le t_n \le T$, we have that

$$\begin{aligned}\sum_{j,k=1}^n a_j a_k C(t_j,t_k) &= \sum_{j,k=1}^n a_j a_k \langle \mathbf{1}_{[0,t_j]}, \mathbf{1}_{[0,t_k]} \rangle_{L^2([0,T],\lambda_1)} \\ &= \left|\left| \sum_{j=1}^n a_j \mathbf{1}_{[0,t_j]} \right|\right|^2_{L^2([0,T],\lambda_1)} \ge 0.\end{aligned}$$

Therefore, function C is both symmetric and positive semi-definite.

DEFINITION 1.10.–*A stochastic process $W = \{W_t, t \ge 0\}$ is called a Wiener process if it is a Gaussian process with $m_t = 0$ and $C(s,t) = s \wedge t$.*

REMARK 1.8.– It follows from example 1.1 and theorem 1.1 that definition 1.10 correctly defines a Wiener process.

PROPOSITION 1.4.– *A Wiener process $W = \{W_t, t \ge 0\}$ starts from zero, has independent increments and $W_t - W_s \sim \mathcal{N}(0, t-s)$.*

PROOF.– First, $\mathbb{E}(W_0)^2 = 0 \wedge 0$; therefore, W starts from zero. Furthermore, its increments are Gaussian and uncorrelated and consequently pair-wise independent if they are taken over non-overlapping intervals. Indeed, for any $0 \le s \le t \le u \le v$ $\mathbb{E}(W_v - W_u)(W_t - W_s) = v \wedge t - u \wedge t - v \wedge s + u \wedge s = 0$. Moreover, for $t > s$ $\mathbb{E}(W_t - W_s)^2 = t - 2s \wedge t + s = t - s$. Therefore, $W_t - W_s \sim \mathcal{N}(0, t-s)$, and the characteristic function of the incremental vector $(W_{t_i} - W_{t_{i-1}}, 1 \leq i \leq n)$ has the form

$$\begin{aligned}\mathbb{E}\exp\left(\sum_{i=1}^{n}\lambda_i(W_{t_i} - W_{t_{i-1}})\right) &= \exp\left(-1/2\sum_{i=1}^{n}\lambda_i^2(t_i - t_{i-1})\right)\\ &= \prod_{i=1}^{n}\exp\left(-1/2\lambda_i^2(t_i - t_{i-1})\right);\end{aligned}$$

therefore, the increments are mutually independent. □

REMARK 1.9.– A Wiener process is also called a Brownian motion.

1.3.3. *Continuity and Hölder property of Gaussian processes*

Consider the case when $\mathbb{T} = [0, T]$. In the following, we consider the separable modifications of all stochastic processes. For the notion of separability, see, for example, [DOO 53] and [MIS 17]. Roughly speaking, separability means that the trajectories of a process are regular in the following sense: a.s. they have the property that the value of a trajectory at any point is situated between the lower and upper bounds of the trajectory at this point. It is a very mild assumption because any stochastic process has a separable modification. Separable modification of a stochastic process together with smoothness of its characteristics provides the smoothness of its trajectories, which will be discussed in detail later. The next result illustrates the above properties of the smoothness of the trajectories of a separable modification of a stochastic process.

THEOREM 1.2.–

1) Let $X = \{X_t, t \in [0, T]\}$ be a stochastic process satisfying the assumption: there exist constants $K > 0$, $\alpha > 0$ and $\beta > 0$ such that

$$\mathbb{E}|X_t - X_s|^\alpha \leq K|t - s|^{1+\beta}, \quad 0 \leq s < t \leq T. \qquad [1.6]$$

Then, X possesses a continuous modification.

2) Moreover, let stochastic process $X = \{X_t, t \in [0, T]\}$ satisfy condition [1.6]. *Then, X possesses a modification with trajectories that are Hölder continuous up to order β/α.*

EXAMPLE 1.2.– *Consider a Wiener process W. Because its increments are Gaussian and for any $0 \leq s \leq t$ $\mathbb{E}(W_t - W_s)^2 = t - s$, then, according to the properties of Gaussian distribution (see exercise 1.3), for any $p \in \mathbb{N}$*

$$\mathbb{E}(W_t - W_s)^{2p} = C_p(t - s)^p,$$

with some constant C_p. If we put $p = 2$, we immediately get from the first statement of theorem 1.2 that W possesses a continuous modification, or, equivalently, its separable modification is a continuous process. Now, let $p \to \infty$. Then, the value

$$\frac{p-1}{2p} = 1/2 - \frac{1}{2p}$$

tends to $1/2$ from below. Again, applying the second statement of theorem 1.2, we get that a Wiener process a.s. has the trajectories which are Hölder up to order $1/2$. It is an exact Hölder property in the sense that the trajectories of a Wiener process are not Hölder of order $1/2$. Moreover, assume that for some Gaussian process X we have

$$\mathbb{E}(X_t - X_s)^2 \leq C(t-s)^{2H},$$

for some $H \in (0,1)$. Then, we can provide the same bounds and conclude that X is Hölder up to order H.

1.3.4. *Uniform bounds and maximal inequalities for Gaussian processes*

We note that in the general case, the exact formulae for the distribution of the maximum of the stochastic process on the interval are unknown. They are unknown even for the majority of Gaussian processes. Only for a Wiener process are the distributions of all natural functionals known, and they are collected in the book [BOR 02], which can be regarded as a kind of encyclopedia. Therefore, much effort was made to find suitable bounds for the distribution of suprema and maxima of Gaussian processes. In this connection, we mention only the books [TAL 05, TAL 14, LIF 12, PIT 96] and the survey [LI 01], and see also the references therein. Fortunately, many properties of Gaussian processes and in particular, uniform bounds and maximal inequalities, are based on such a simple characteristic as the square incremental distance. It is defined for a stochastic process $X = \{X_t, t \in \mathbb{T}\}$ as

$$d_X(s,t) = \left(\mathbb{E}(X_s - X_t)^2\right)^{1/2}, s,t \in \mathbb{T}.$$

REMARK 1.10.– Function $d_X(\cdot,\cdot)$ has all properties of the semi-metric (see remark A.5), and, if it follows from $d_X(s,t) = 0$ that $s = t$, it becomes a metric (see definition A.10). Assume that the process X is pair-wise non-degenerate, i.e. any pair $(X_s, X_t), s \neq t$, is linearly independent. Then, $X_s - X_t, s \neq t$, is a non-degenerate Gaussian random variable; therefore, $d_X(s,t) > 0$ for any $s \neq t$, and consequently, $d_X(\cdot,\cdot)$ is a metric. In the following, we shall assume that this condition is fulfilled.

1.3.4.1. *Uniform maximal bound in probability for a Gaussian process*

The first result is a uniform bound in probability for a Gaussian process defined on some interval $\mathbb{T} = [0, T] \subset \mathbb{R}$.

Let $X = \{X_t, t \in \mathbb{T}\}$ be a Gaussian process, $\rho(\varepsilon) = \sup_{s,t \in \mathbb{T}, |s-t| \leq \varepsilon} d_X(s,t)$ and $Q(\delta) = \int_0^\infty \rho(\delta e^{-y^2}) dy$. Assume that ρ is strictly increasing and denote $Q^{-1}(\delta)$ the inverse function. Let $\sigma^2 = \sup_{t \in \mathbb{T}} \mathbb{E} X_t^2$. The following theorem is formulated in [LI 01], where it is emphasized that the bound is sharp, according to [BER 85].

THEOREM 1.3.– *For all $x > 0$*

$$\mathbb{P}\{\sup_{t \in \mathbb{T}} |X_t| > x\} \leq C(Q^{-1}(1/x))^{-1} \exp\left\{-\frac{x^2}{2\sigma^2}\right\}.$$

1.3.4.2. *Expectation maximal bound for a Gaussian process and its increments in terms of entropy*

Consider a Gaussian process $X = \{X_t, t \in \mathbb{T}\}$ and the respective distance $d_X(s,t) = \left(\mathbb{E}(X_t - X_s)^2\right)^{1/2}$. Let D be the diameter of $\mathbb{T}$ with respect to semi-metric d_X, i.e.

$$D = \sup_{s,t \in \mathbb{T}} d_X(s,t).$$

Denote by $N(\epsilon, d_X, \mathbb{T})$ the minimal number of closed balls of radius $\epsilon > 0$ in the metric d_X with centers in $\mathbb{T}$, needed to cover $\mathbb{T}$. Logarithm of N is often called an ε-entropy, or a metric ε-entropy, of the set $\mathbb{T}$ in metric d_X.

REMARK 1.11.– Again, consider the relation between pair-wise non-degeneracy of the process X and its metric entropy. Assume for the moment that we omitted the assumption of pair-wise non-degeneracy, and for some $\varepsilon > 0$, there are no closed balls of radius $\epsilon > 0$ in the semi-metric d_X with centers in $\mathbb{T}$, needed to cover $\mathbb{T}$. For example, let the whole process X consist of one Gaussian random variable, $X_t = X$ for all $t \in \mathbb{T}$, or consider a slightly less trivial example, in which there exists a Gaussian vector (X, Y), and $X_t = X$ for all $t \in \mathbb{T}$ except one point, t_0, say, while $X_{t_0} = Y$, and $\left(\mathbb{E}(X - Y)^2\right)^{1/2} = 1$. Then, $N(\epsilon, d_X, \mathbb{T})$ is not defined for any $\epsilon > 0$ in the first case, and for any $0 < \epsilon < 1$ in the second case. For such $\epsilon > 0$, it is natural to put $N(\epsilon, d_X, \mathbb{T}) = 1$ in order to get zero ε-entropy: $\log N(\epsilon, d_X, \mathbb{T}) = 0$. Of course, in this case, we cannot apply the notion of entropy in order to get reasonable upper maximal bounds for the process X. Therefore, in the next theorem, we assume that the process X is pair-wise non-degenerate and d_X is a metric. Moreover, in considering integrals of the form $\int_0^D \sqrt{\log N(\epsilon, d_X, \mathbb{T})} d\epsilon$ (these are called Dudley integrals), it is implicitly assumed that the process X is continuous in the mean-square sense so that $d_X(s,t) \to 0$ as $t - s \to 0$. In this case, $N(\epsilon, d_X, \mathbb{T}) \to \infty$ as $\varepsilon \to 0$ and the question of convergence of Dudley integrals appears.

The proof of the following theorem can be found, for example, in [MAR 06], theorem 6.1.2. We recall that we consider a separable modification of any stochastic process.

THEOREM 1.4.– *Let a Gaussian process X be separable, pair-wise non-degenerate and continuous in the mean-square sense. If $\int_0^D \sqrt{\log N(\epsilon, d_X, \mathbb{T})}d\epsilon < \infty$, then X has bounded uniformly continuous sample paths on $(\mathbb{T}, d_X)$ such that*

i) $\mathbb{E}(\sup_{t\in\mathbb{T}} |X(t)|) \leq C \int_0^{D/2} \sqrt{\log N(\epsilon, d_X, \mathbb{T})}d\epsilon$, *and*

ii) $\mathbb{E}\Big(\sup_{s,t\in\mathbb{T}; d_X(s,t)\leq\delta} |X_t - X_s|\Big) \leq C \int_0^{\delta} \sqrt{\log N(\epsilon, d_X, \mathbb{T})}d\epsilon$,

where C denotes a positive universal constant.

REMARK 1.12.– Assume that the metric d_X can be majorized by the metric d in the sense that for any $s, t \in \mathbb{T}$ $d_X(s,t) \leq d(s,t)$. Then, for any $\epsilon > 0$, $N(\epsilon, d_X, \mathbb{T}) \leq N(\epsilon, d, \mathbb{T})$. It can be easily explained in such a way: let, for example, $\mathbb{T} = [0, T]$, and we cover this interval by closed balls of radius ϵ (simply, by the intervals) in the metric d_X and in metric d, starting from zero. Then, the right end-point, say, t_X, of the first interval in metric d_X, satisfies the equality $d_X(0, t_X) = 2\epsilon$, and for metric d, the right end-point, say, t, of the first interval, satisfies the equality $d(0,t) = 2\epsilon$, whence $t \leq t_X$, and the same order is preserved for the subsequent points; therefore, the minimal number of intervals covering $\mathbb{T}$ in metric d exceeds the respective number in metric d_X.

Consider the following two examples.

EXAMPLE 1.3.– *Let $\mathbb{T} = [0, T]$, and let there exist two constants, $C > 0$ and $\alpha \in (0,1)$, such that $d_X(s,t) \leq C|s-t|^\alpha$. We note that $d(s,t) := C|s-t|^\alpha$ is a metric on $\mathbb{T}$. We have in this metric that $N(\epsilon, d, \mathbb{T}) = \left[\frac{TC^{1/\alpha}}{(2\epsilon)^{1/\alpha}}\right] + 1$, where $[x]$ is the greatest integer not exceeding x.*

EXAMPLE 1.4.– *Let $\mathbb{T} = [0, T]$; processes X and Y are independent zero mean (i.e. centered) pair-wise non-degenerate Gaussian processes. Then, their sum generates metric $d_{X+Y} = \sqrt{d_X^2 + d_Y^2}$, and $d_{X+Y} \geq d_X \vee d_Y$. Consequently,*

$$N(\epsilon, d_X, \mathbb{T}) \vee N(\epsilon, d_Y, \mathbb{T}) \leq N(\epsilon, d_{X+Y}, \mathbb{T}).$$

This inequality is completely consistent with the physical concept of entropy as a measure of the chaos of a dynamic system. Namely, the more independent processes participate in the dynamic system, thereby creating its additional degrees of freedom, the more chaos exists in the system and thus its entropy is greater.

1.3.4.3. *Asymptotic behavior of maximal expectation functionals of Gaussian processes*

Consider the lower and upper bounds of the expected maximum of a Gaussian process $X = \{X_t, t \in \mathbb{T}\}$ for which the distance $d_X(s,t)$ admits two-side power bounds, similar to those described in example 1.3. Assume that $d_X(s,t)$ satisfies the following condition:

i) There exist two constants $0 \le C_1 \le C_2$ and a constant $H \in (0,1)$ such that

$$C_1|t-s|^H \le d_X(s,t) \le C_2|t-s|^H \qquad [1.7]$$

for any $t, s \in [0,1]$.

DEFINITION 1.11.– *For the Gaussian process satisfying inequalities* [1.7] *from condition* (i), *we say that it belongs to quasi-helix*

$$\mathbb{QH}(H, C_1, C_2, \mathbb{T}),$$

according to terminology introduced in [KAH 85].

As was mentioned above, such a process is pair-wise non-degenerate and d_X is a metric. Consider now a Gaussian process $X = \{X_t, t \in [0,1]\}$ defined on the fixed interval, which is supposed to be $[0,1]$ for technical simplicity. In the following theorem, the left- and right-hand sides of inequality [1.7] are considered separately. The next results were proved in [BOR 17].

THEOREM 1.5.–

i) If there exist $C_1 > 0$ *and* $H \in (0,1)$ *such that* $d_X(s,t) \ge C_1|t-s|^H$ *for any* $t, s \in [0,1]$, *then*

$$\mathbb{E} \max_{0 \le t \le T} X_t \ge \frac{C_1}{5\sqrt{H}}.$$

ii) If there exist $C_2 > 0$ *and* $H \in (0,1)$ *such that* $d_X(s,t) \le C_2|t-s|^H$ *for any* $t, s \in [0,1]$, *then*

$$\mathbb{E} \max_{0 \le t \le 1} X_t \le \frac{16.3 C_2}{\sqrt{H}}.$$

1.3.5. *Equivalence and singularity of measures related to Gaussian processes*

1.3.5.1. *Cylinder sets*

Consider a probability space $(\Omega, \mathcal{F}, \mathbb{P})$, a non-empty set $\mathbb{T}$, the set $\mathbb{R}^{\mathbb{T}}$ of the real-valued functions on $\mathbb{T}$ and the set $C(\mathbb{T})$ of continuous real-valued functions on $\mathbb{T}$.

DEFINITION 1.12.– *A subset A of $\mathbb{R}^{\mathbb{T}}$ (respectively, $C(\mathbb{T})$) is called a cylinder set in $\mathbb{R}^{\mathbb{T}}$ (in $C(\mathbb{T})$), if there exist an integer $n \in \mathbb{N}$, $t_j \in \mathbb{T}$ for $j = 1, \ldots, n$ and a Borel subset $B \subset \mathbb{R}^n$ such that*

$$A = \left\{x \in \mathbb{R}^{\mathbb{T}} (\textit{respectively}, C(\mathbb{T})) | \, (x(t_1), x(t_2), \ldots, x(t_n)) \in B\right\}.$$

The smallest σ-field containing all cylinder sets of $\mathbb{R}^{\mathbb{T}}$ (respectively, $C(\mathbb{T})$) is denoted by $\mathcal{F}^{\mathbb{T}}$ (respectively, $\mathcal{C}^{\mathbb{T}}$).

1.3.5.2. *Probability measure induced by a stochastic process*

DEFINITION 1.13.– *Consider a stochastic process $X = \{X_t, t \in \mathbb{T}\}$ defined on the probability space $(\Omega, \mathcal{F}, \mathbb{P})$. We introduce a set function $\mathbb{P}_X$ on a measurable space $\left(\mathbb{R}^{\mathbb{T}}, \mathcal{F}^{\mathbb{T}}\right)$ such that*

$$\mathbb{P}_X \left(x \in \mathbb{R}^{\mathbb{T}} | \, (x(t_1), \ldots, x(t_n)) \in B\right) = \mathbb{P}\left(\omega \in \Omega | \, (X_{t_1}(\omega), \ldots, X_{t_n}(\omega)) \in B\right),$$

for any $t_1, \ldots, t_n \in \mathbb{T}$ and for any Borel subset $B \subset \mathbb{R}^n$.

REMARK 1.13.– It is possible to prove that $\mathbb{P}_X$ is indeed a probability measure on a measurable space $\left(\mathbb{R}^{\mathbb{T}}, \mathcal{F}^{\mathbb{T}}\right)$, see, for example, [MIS 17]. The probability measure $\mathbb{P}_X$ is called the measure induced by the process X.

1.3.5.3. *Gaussian measures: entropy and entropy distance for two measures*

DEFINITION 1.14.– *A probability measure $\mathbb{P}_X$ on $\left(\mathbb{R}^{\mathbb{T}}, \mathcal{F}^{\mathbb{T}}\right)$ is said to be Gaussian if it is induced by a Gaussian process $X = \{X_t, t \in \mathbb{T}\}$. If, in particular, the measure $\mathbb{P}_X$ is supported by the space $C(\mathbb{T})$ of continuous functions on $\mathbb{T}$, then $\mathbb{P}_X$ is called a Gaussian measure on $C(\mathbb{T})$.*

DEFINITION 1.15.– *Consider two probability measures $\mathbb{P}_k$, $k = 1, 2$, on a measurable space $(\Omega, \mathcal{F})$, and let $\mathcal{P} = \{A_i \in \mathcal{F}, i = 1, \ldots, n, n \geq 1\}$ be a class of all possible finite partitions of Ω. Then, the value*

$$\mathbf{H}(\mathbb{P}_2|\mathbb{P}_1) = \sup_{\mathcal{P}} \sum_{i=1}^{n} \mathbb{P}_2(A_i) \log \frac{\mathbb{P}_2(A_i)}{\mathbb{P}_1(A_i)}$$

is called the relative entropy of the probability measure $\mathbb{P}_2$ relative to $\mathbb{P}_1$. Note that in the formula for $\mathbf{H}(\mathbb{P}_2|\mathbb{P}_1)$, we assume that $0 \log 0 = 0$ and $\log 0 = -\infty$.

DEFINITION 1.16.– *The entropy distance $\mathbf{I}(\mathbb{P}_1, \mathbb{P}_2)$ between two measures $\mathbb{P}_1$ and $\mathbb{P}_2$ is defined by*

$$\mathbf{I}(\mathbb{P}_1, \mathbb{P}_2) = \mathbf{H}(\mathbb{P}_2|\mathbb{P}_1) + \mathbf{H}(\mathbb{P}_1|\mathbb{P}_2).$$

REMARK 1.14.– Because the function $F(x) = x \log x$ is convex, we can apply Jensen's inequality to conclude that

$$\mathbf{H}(\mathbb{P}_2|\mathbb{P}_1) = \sup_{\mathcal{P}} \sum_{i=1}^{n} \mathbb{P}_2(A_i) \log \frac{\mathbb{P}_2(A_i)}{\mathbb{P}_1(A_i)} \geq \sup_{\mathcal{P}} \sum_{i=1}^{n} \mathbb{P}_1(A_i) \frac{\mathbb{P}_2(A_i)}{\mathbb{P}_1(A_i)} \log \frac{\mathbb{P}_2(A_i)}{\mathbb{P}_1(A_i)}$$

$$\geq \sup_{\mathcal{P}} \left(\sum_{i=1}^{n} \mathbb{P}_1(A_i) \frac{\mathbb{P}_2(A_i)}{\mathbb{P}_1(A_i)} \right) \log \left(\sum_{i=1}^{n} \mathbb{P}_1(A_i) \frac{\mathbb{P}_2(A_i)}{\mathbb{P}_1(A_i)} \right) = 1 \log 1 = 0.$$

This means that the relative entropy is non-negative.

REMARK 1.15.– Recall the notions of absolutely continuous, equivalent and singular probability measures. We say that the probability measure $\mathbb{Q}$ is absolutely continuous w.r.t. the measure $\mathbb{P}$, both defined on a probability space $(\Omega, \mathcal{F})$ (denoted by $\mathbb{Q} \ll \mathbb{P}$), if for any set $A \in \mathcal{F}$ such that $\mathbb{P}(A) = 0$ we have that $\mathbb{Q}(A) = 0$. If $\mathbb{Q} \ll \mathbb{P}$, then there exists a non-negative random variable (Radon–Nikodym derivative, Radon–Nikodym density) $\frac{d\mathbb{Q}}{d\mathbb{P}}$ such that for any set $A \in \mathcal{F}$

$$\mathbb{Q}(A) = \int_{\Omega} \frac{d\mathbb{Q}}{d\mathbb{P}} d\mathbb{P}(\omega).$$

We say that the probability measures $\mathbb{Q}$ and $\mathbb{P}$ are equivalent, $\mathbb{Q} \sim \mathbb{P}$, if both $\mathbb{Q} \ll \mathbb{P}$ and $\mathbb{P} \ll \mathbb{Q}$. In this case, there exist both $\frac{d\mathbb{Q}}{d\mathbb{P}}$ and $\frac{d\mathbb{P}}{d\mathbb{Q}}$, both of which are positive a.s., and moreover,

$$\frac{d\mathbb{Q}}{d\mathbb{P}} \cdot \frac{d\mathbb{P}}{d\mathbb{Q}} = 1.$$

We say that the probability measures $\mathbb{Q}$ and $\mathbb{P}$ are singular if there exists such a set $A \in \mathcal{F}$ that $\mathbb{Q}(A) = 0$ and $\mathbb{P}(A) = 1$ (in this case, obviously, there exists a set $B = \Omega \setminus A \in \mathcal{F}$ such that $\mathbb{Q}(B) = 1$ and $\mathbb{P}(B) = 0$, therefore the definition of singularity is symmetric). If two measures $\mathbb{P}_1$ and $\mathbb{P}_2$ are singular, then obviously $\mathbf{I}(\mathbb{P}_1, \mathbb{P}_2) = \infty$. However, we cannot state that if $\mathbf{I}(\mathbb{P}_1, \mathbb{P}_2) = \infty$, then the measures $\mathbb{P}_1$ and $\mathbb{P}_2$ are singular. Conversely, they can even be equivalent. To clarify this situation, let us take any positive random variable Z such that $\mathbb{E}Z = 1$. Then, Z can be considered as a Radon–Nikodym derivative, $Z = \frac{d\mathbb{Q}}{d\mathbb{P}}$ for some probability measure $\mathbb{Q} \sim \mathbb{P}$. Assume that $\mathbb{E}Z \log Z = \infty$, and we immediately get an example of equivalent probability measures with infinite relative entropy and entropy distance.

LEMMA 1.9.– *Let $\mathbb{P}_1$ and $\mathbb{P}_2$ be two probability measures on $(\Omega, \mathcal{F})$, and let there exist some $\varepsilon_0 \in (0, 1]$ such that for any $\varepsilon \leq \varepsilon_0$, in turn, there exists an event A_ε such that $\mathbb{P}_1(A_\varepsilon) \leq \varepsilon$ and $\mathbb{P}_2(A_\varepsilon) \geq 1 - \varepsilon$. Then, the measures $\mathbb{P}_1$ and $\mathbb{P}_2$ are singular.*

PROOF.– Put $\varepsilon_k = \frac{1}{k^2}$, and denote $B_k = A_{\varepsilon_k}$. Consider the event

$$A = \bigcap_{n=1}^{\infty} \bigcup_{k=n}^{\infty} B_k.$$

Then, on the one hand, for any $n \geq 1$

$$\mathbb{P}_2 \left(\bigcup_{k=n}^{\infty} B_k \right) \geq \sup_{k \geq n} \mathbb{P}_2(B_k) = 1,$$

whence it follows immediately that

$$\mathbb{P}_2(A) = \lim_{n \to \infty} \mathbb{P}_2 \left(\bigcup_{k=n}^{\infty} B_k \right) = 1.$$

On the other hand, the events $\bigcup_{k=n}^{\infty} B_k$ are decreasing, therefore

$$\mathbb{P}_1(A) = \lim_{n \to \infty} \mathbb{P}_1 \left(\bigcup_{k=n}^{\infty} B_k \right) \leq \lim_{n \to \infty} \sum_{k=n}^{\infty} \mathbb{P}_1(B_k) \leq \lim_{n \to \infty} \sum_{k=n}^{\infty} \frac{1}{k^2} = 0,$$

and a lemma is proved. □

The proof of the following lemmas can be found, for example, in [HID 93], Chapter 6, section 2.

LEMMA 1.10.–

1) If a probability measure $\mathbb{P}_2$ is absolutely continuous with respect to a probability measure $\mathbb{P}_1$, then the entropy $\mathbf{H}(\mathbb{P}_2|\mathbb{P}_1)$ can be expressed by using the Radon–Nikodym density $\varphi(\omega) = \dfrac{d\mathbb{P}_2(\omega)}{d\mathbb{P}_1(\omega)}$ as follows

$$\mathbf{H}(\mathbb{P}_2|\mathbb{P}_1) = \mathbb{E}\left(\varphi(\omega)\log(\varphi(\omega))\right) = \int_{\Omega} \log\varphi(\omega) d\mathbb{P}_2(\omega)$$
$$= \int_{\Omega} (\log\varphi(\omega))\varphi(\omega) d\mathbb{P}_1(\omega).$$

2) If $\mathbb{P}_2$ is not absolutely continuous with respect to $\mathbb{P}_1$, then $\mathbf{H}(\mathbb{P}_2|\mathbb{P}_1) = \infty$.

LEMMA 1.11.– *Let $\{X_k, 1 \leq k \leq n\}$ be a finite number of random variables on a measurable space $(\Omega, \mathcal{F})$. Assume that X_k are independent and Gaussian with respect to both probability measures $\mathbb{P}_1$ and $\mathbb{P}_2$, and let*

$$\mathbb{E}_1(X_k) = m_k,\ \mathrm{Var}_1(X_k) = \sigma_k^2 > 0,\ \mathbb{E}_2(X_k) = 0,\ \mathrm{Var}_2(X_k) = 1,\ 1 \leq k \leq n,$$

where $\mathbb{E}_i$ and Var_i are the mean and variance with respect to $\mathbb{P}_i, i = 1, 2$, respectively. Let $\mathcal{G}$ be the sub-σ-algebra generated by the sequence $\{X_i, 1 \leq i \leq n\}$. Then, the restrictions $\mathbb{P}_1'$ and $\mathbb{P}_2'$ of $\mathbb{P}_1$ and $\mathbb{P}_2$, respectively, to $\mathcal{G}$ are equivalent probability measures and the following relations hold:

i) $$\log\left(\frac{d\mathbb{P}_2'(\omega)}{d\mathbb{P}_1'(\omega)}\right) = \log\varphi'(\omega) = \frac{1}{2}\sum_{k=1}^{n}\left(\log\sigma_k^2 + \frac{(X_k - m_k)^2}{\sigma_k^2} - X_k^2\right).$$

ii) $$\mathbb{E}_1(\log\varphi') = -\frac{1}{2}\sum_{k=1}^{n}\left(\sigma_k^2 - 1 - \log\sigma_k^2 + m_k^2\right).$$

iii) $$\mathbb{E}_2(\log\varphi') = \frac{1}{2}\sum_{k=1}^{n}\left(\sigma_k^{-2} - 1 + \log\sigma_k^2 + m_k^2\sigma_k^{-2}\right).$$

iv) $$\mathrm{Var}_1(\log\varphi') = \frac{1}{2}\sum_{k=1}^{n}\left((\sigma_k^2 - 1)^2 + 2m_k^2\sigma_k^2\right).$$

v) $$\mathrm{Var}_2(\log\varphi') = \frac{1}{4}\sum_{k=1}^{n}\left((\sigma_k^2 - 1)^2\sigma_k^{-4} + 2m_k^2\sigma_k^{-4}\right).$$

Now, we consider a process $X = \{X_t, t \in \mathbb{T}\}$ that is Gaussian with respect to both probability measures $\mathbb{P}_1$ and $\mathbb{P}_2$. To be specific, we take the basic measurable space $(\Omega, \mathcal{F})$ to be $\left(\mathbb{R}^{\mathbb{T}}, \mathcal{F}^{\mathbb{T}}\right)$ on which two Gaussian measures $\mathbb{P}_1$ and $\mathbb{P}_2$ are given. Let $\mathcal{G}_{\mathbb{T}'}$ be the sub-σ-field of $\mathcal{F}$ generated by $\{X_t, t \in \mathbb{T}'\}$, where $\mathbb{T}'$ is a finite subset of $\mathbb{T}$.

The proof of the following lemma can be found, for example, in [HID 93], Chapter 6, section 2.

LEMMA 1.12.– *Suppose that an increasing sequence $\{\mathcal{F}_n, n \geq 1\}$ of sub-σ-fields converges to $\mathcal{F}$ so that $\mathcal{F} = \vee_n \mathcal{F}_n$. Let $\mathbb{P}_1^n$ and $\mathbb{P}_2^n$ be the restrictions of the given probability measures $\mathbb{P}_1$ and $\mathbb{P}_2$ to $\mathcal{F}_n$, respectively. If*

$$\sup_n \mathbf{H}(\mathbb{P}_2^n | \mathbb{P}_1^n) < \infty,$$

then $\mathbb{P}_2$ is absolutely continuous with respect to $\mathbb{P}_1$ and

$$\mathbf{H}(\mathbb{P}_2 | \mathbb{P}_1) = \sup_n \mathbf{H}(\mathbb{P}_2^n | \mathbb{P}_1^n).$$

DEFINITION 1.17.– *Let a process* $X = \{X_t, t \in \mathbb{T}\}$ *be Gaussian with respect to the measure* $\mathbb{P}$. *We say that the process* X *is non-degenerate w.r.t. this measure if, for any* $n \geq 1$ *and any* $t_1, t_2, \ldots, t_n \in \mathbb{T}$ *covariance matrix* $\big(\mathrm{Cov}(X_{t_i}, X_{t_j}), 1 \leq i, j \leq n\big)$ *is positive definite.*

The next result can be proved similarly to lemma 1.12.

LEMMA 1.13.– *Entropy distance between two measures* $\mathbb{P}_1$ *and* $\mathbb{P}_2$ *equals*

$$\mathbf{I}(\mathbb{P}_1, \mathbb{P}_2) = \sup_{\mathbb{T}' \in \Delta} \mathbf{I}(\mathbb{P}_1', \mathbb{P}_2'),$$

where $\mathbf{I}(\mathbb{P}_1', \mathbb{P}_2') \equiv \mathbf{H}(\mathbb{P}_1' | \mathbb{P}_2') + \mathbf{H}(\mathbb{P}_2' | \mathbb{P}_1')$, Δ *is the class of all finite subsets* $\mathbb{T}'$ *of* $\mathbb{T}$, *and* $\mathbb{P}_1'$ *(respectively* $\mathbb{P}_2'$*) are the restrictions of* $\mathbb{P}_1$ *(respectively* $\mathbb{P}_2$*) to* $\mathcal{G}_{\mathbb{T}'}$.

1.3.5.4. *Equivalency and singularity of Gaussian measures*

From now on, we assume that the set $\mathbb{T}$ is infinite.

DEFINITION 1.18.– *Let a process* $X = \{X_t, t \in \mathbb{T}\}$ *be Gaussian with respect to the measure* $\mathbb{P}$. *We say that the process* X *is non-degenerate w.r.t. this measure if for any* $n \geq 1$ *and any* $t_1, t_2, \ldots, t_n \in \mathbb{T}$ *covariance matrix* $\big(\mathrm{Cov}(X_{t_i}, X_{t_j}), 1 \leq i, j \leq n\big)$ *is positive definite. In the opposite case, we say that the process* X *is degenerate w.r.t. the measure* $\mathbb{P}$.

Now our goal is to establish singularity conditions for two Gaussian measures corresponding to the same process. Consider the process $X = \{X_t, t \in \mathbb{T}\}$ and assume that it is Gaussian with respect to two probability measures $\mathbb{P}_1$ and $\mathbb{P}_2$. Take any finite number of points $t_1, t_2, \ldots, t_n$ from $\mathbb{T}$ and consider vector $X = (X_{t_1}, X_{t_2}, \ldots, X_{t_n})$ which is a Gaussian vector with respect to both measures. Denote $K_{X,1}$ and $K_{X,2}$ its covariance matrices w.r.t. $\mathbb{P}_1$ and $\mathbb{P}_2$, respectively. According to theorem A.8, there exists a non-degenerate matrix S such that both $S^T K_{X,1} S$ and $S^T K_{X,2} S$ are diagonal matrices. Moreover, according to lemma 1.8, $S^T K_{X,1} S$ and $S^T K_{X,2} S$ are covariance matrices of the vector SX with respect to the corresponding measures. In this connection, let us consider two cases.

i) Let there exist such a vector X that matrices $S^T K_{X,1} S$ and $S^T K_{X,2} S$ have different numbers of non-zero diagonal elements. This means that vector SX has a different number of independent components w.r.t. measures $\mathbb{P}_1$ and $\mathbb{P}_2$. Let them be $1 \leq k_1 < k_2 \leq n$. Then, there exists a random vector $Y = (Y_1, Y_2, \ldots, Y_{k_2})$ whose components are taken from SX and non-random non-zero vector $\lambda = (\lambda_1, \lambda_2, \ldots, \lambda_{k_2})$ such that the linear combination $\sum_{k=1}^{k_2} \lambda_k Y_k$ has a non-degenerate Gaussian distribution w.r.t. $\mathbb{P}_2$ and a degenerate distribution w.r.t. $\mathbb{P}_1$, i.e.

$$\mathbb{P}_1\left(\sum_{k=1}^{k_2} \lambda_k Y_k = 0\right) = 1,$$

and

$$\mathbb{P}_2\left(\sum_{k=1}^{k_2}\lambda_k Y_k = 0\right) = 0.$$

This means that the measures $\mathbb{P}_1$ and $\mathbb{P}_2$ are singular. As was mentioned before, in this case, their entropy distance $\mathbf{I}(\mathbb{P}_1, \mathbb{P}_2) = \infty$. Of course, this is also true in the case when the process X is non-degenerate w.r.t. one of the measures and degenerate w.r.t. another.

ii) Let, for any vector X, matrices $S^T K_{X,1} S$ and $S^T K_{X,2} S$ have the same number of non-zero diagonal elements. This means that vector SX has the same number of independent components w.r.t. both measures. Assume that the process X is degenerate in the sense that the number of such components is restricted by some number k. Consider any vector X with k linearly independent components. Then, both matrices $K_{X,1}$ and $K_{X,2}$ are symmetric and positive definite; therefore, we can apply theorem A.9 and conclude that the restrictions of the measures $\mathbb{P}_1$ and $\mathbb{P}_2$ on this vector are both non-degenerate. Then, we can apply lemma 1.11 and conclude that the measures $\mathbb{P}_1$ and $\mathbb{P}_2$, being concentrated on the vector X, are equivalent.

From now on, we can exclude both cases (i) and (ii) considered in detail above and concentrate on non-degenerate Gaussian processes. Recall that this means that any vector $(X_{t_1}, \dots, X_{t_n})$ consists of linearly independent components and consequently has a non-degenerate Gaussian distribution w.r.t. two probability measures, $\mathbb{P}_1$ and $\mathbb{P}_2$, on $(\Omega, \mathcal{F})$. The idea of the next statement was taken from [HID 93], Chapter 6; however, we present this proof with slight modifications, for the reader's benefit.

THEOREM 1.6.– *Let the process $X = \{X_t, t \in \mathbb{T}\}$ be Gaussian with respect to two measures $\mathbb{P}_1$ and $\mathbb{P}_2$ on $(\Omega, \mathcal{F})$ and non-degenerate w.r.t. both measures. If the entropy distance $\mathbf{I}(\mathbb{P}_1, \mathbb{P}_2)$ is infinite, then $\mathbb{P}_1$ and $\mathbb{P}_2$ are singular.*

PROOF.– Without loss of generality, assume that $\mathbb{E}_2(X_t) = 0, t \in \mathbb{T}$. Indeed, if this is not the case, then we can take $\tilde{X}_t = X_t - \mathbb{E}_2(X_t)$ in place of X_t. Then, the covariance matrices are the same, the linear independence of the components is preserved and further proof is carried out in the same way as without this condition.

For any finite subset $\mathbb{T}' = \{t_1, \dots, t_n\}$, denote as before

$$K_{X,1} = \Big(\mathrm{Cov}_1(X_{t_i}, X_{t_j})\Big)_{1\le i,j\le n} \text{ and } K_{X,2} = \Big(\mathrm{Cov}_2(X_{t_i}, X_{t_j})\Big)_{1\le i,j\le n},$$

the covariance matrices of the vector $(X_{t_1}, \dots, X_{t_n})$ with respect to $\mathbb{P}_1$ and $\mathbb{P}_2$, respectively. According to our supposition, both covariance matrices are positive definite. Then, it follows from theorem A.9 that there exists an invertible matrix

$C = (C_{ij})_{1 \le i,j \le n}$ such that $C^T K_{X,1} C$ and $C^T K_{X,2} C$ are simultaneously diagonal and one of them, say $C^T K_{X,2} C$, is the unit matrix. Set

$$Y_k = \sum_{j=1}^{n} C_{kj} X_{t_j}, \; k = 1, 2, \ldots, n.$$

Then, $Y = \{Y_k, 1 \le k \le n\}$ is a Gaussian sequence consisting of independent, w.r.t. both measures $\mathbb{P}_1$ and $\mathbb{P}_2$, random variables, with

$$\mathbb{E}_2(Y_k Y_j) = (C^T K_{X,2} C)_{k,j} = \begin{cases} 1 \text{ if } k = j \\ 0 \text{ if } k \ne j. \end{cases}$$

Set $\mathcal{G} = \sigma(Y) = \sigma(X_{t_1}, \ldots, X_{t_n})$. Denoting by $\mathbb{P}_1'$ and $\mathbb{P}_2'$ the restrictions of $\mathbb{P}_1$ and $\mathbb{P}_2$ to $\mathcal{G}$, respectively, and $\varphi'(\omega) = \frac{d\mathbb{P}_2'(\omega)}{d\mathbb{P}_1'(\omega)}$, we obtain from lemmas 1.10 and 1.11 that

$$\begin{aligned} \mathbf{I}(\mathbb{P}_1', \mathbb{P}_2') &= \mathbf{H}(\mathbb{P}_1' | \mathbb{P}_2') + \mathbf{H}(\mathbb{P}_2' | \mathbb{P}_1') \\ &= \int_\Omega \log\Big(\varphi'(\omega)\Big) \mathbb{P}_2'(d\omega) + \int_\Omega \log\Big((\varphi'(\omega))^{-1}\Big) \mathbb{P}_1'(d\omega) \\ &= \mathbb{E}_2\Bigg(\log \varphi'(\omega)\Bigg) - \mathbb{E}_1\Bigg(\log \varphi'(\omega)\Bigg) \qquad [1.8] \\ &= \frac{1}{2} \sum_{k=1}^{n} \left(\frac{(\sigma_k - 1)^2}{\sigma_k^2} + m_k^2 \left(1 + \frac{1}{\sigma_k^2}\right) \right), \end{aligned}$$

where $m_k = \mathbb{E}_1(Y_k), \sigma_k^2 = \mathrm{Var}_1(Y_k)$. To prove that the measures $\mathbb{P}_1$ and $\mathbb{P}_2$ are singular, we consider two cases.

i) Let there exist two positive constants c_1 and c_2 such that

$$c_1 \le \sigma_k \le c_2, \; 1 \le k \le n,$$

for any finite subset $\mathbb{T}' = \{t_1, \ldots, t_n\}$ of $\mathbb{T}$. It follows from equality [1.8] that there exist positive constants a and b such that

$$a \sum_{k=1}^{n} \left((\sigma_k^2 - 1)^2 + m_k^2 \right) \le \mathbf{I}(\mathbb{P}_1', \mathbb{P}_2') \le b \sum_{k=1}^{n} \left((\sigma_k^2 - 1)^2 + m_k^2 \right), \qquad [1.9]$$

$$\mathrm{Var}_1(\log \varphi') \le b \sum_{k=1}^{n} \left((\sigma_k^2 - 1)^2 + m_k^2 \right) \qquad [1.10]$$

and

$$\mathrm{Var}_2(\log\varphi') \le b\sum_{k=1}^{n}\left((\sigma_k^2-1)^2+m_k^2\right), \qquad [1.11]$$

hold. Denote by A' the event in $\mathcal{G}$ defined by

$$A' = \left\{\omega\in\Omega : -\log\varphi'(\omega) > \frac{1}{2}\mathbf{I}(\mathbb{P}_1',\mathbb{P}_2') - \mathbb{E}_2(\log\varphi')\right\}. \qquad [1.12]$$

It follows from Chebyshev's inequality that

$$\mathbb{P}_2(A') = \mathbb{P}_2\left(\omega\in\Omega : -\log\varphi'(\omega) + \mathbb{E}_2(\log\varphi') > \frac{1}{2}\mathbf{I}(\mathbb{P}_1',\mathbb{P}_2')\right)$$
$$\le 4\mathrm{Var}_2(\log\varphi')\,(\mathbf{I}(\mathbb{P}_1',\mathbb{P}_2'))^{-2} \le c\mathbf{I}(\mathbb{P}_1',\mathbb{P}_2')\Big)^{-1},$$

with $c = 4\frac{b}{a}$, where in the last inequality we used equations [1.9], [1.10] and [1.11].

Moreover, it follows from equation [1.8] that

$$A' = \left\{\omega\in\Omega| -\log\varphi'(\omega) > \frac{1}{2}\mathbf{I}(\mathbb{P}_1',\mathbb{P}_2') - \mathbb{E}_2(\log\varphi')\right\}$$
$$= \left\{\omega\in\Omega| -\log\varphi'(\omega) > -\frac{1}{2}\mathbf{I}(\mathbb{P}_1',\mathbb{P}_2') - \mathbb{E}_1(\log\varphi')\right\}.$$

Thus

$$\mathbb{P}_1(\Omega\setminus A') = \mathbb{P}_1\Big(\omega\in\Omega|\log\varphi' - \mathbb{E}_1(\log\varphi') \ge \frac{1}{2}\mathbf{I}(\mathbb{P}_1',\mathbb{P}_2')\Big).$$

Again, it follows from Chebyshev's inequality together with relations [1.9], [1.10] and [1.11] that

$$\mathbb{P}_1(A') = 1 - \mathbb{P}_1'(\Omega\setminus A') \ge 1 - \mathrm{Var}_1(\log\varphi')\Big(\frac{1}{2}\mathbf{I}(\mathbb{P}_1',\mathbb{P}_2')\Big)^{-2}$$
$$\ge 1 - c\,\mathbf{I}(\mathbb{P}_1',\mathbb{P}_2')\Big)^{-1},$$

with the same constant with $c = 4\frac{b}{a}$.

From the assertion $\mathbf{I}(\mathbb{P}_1,\mathbb{P}_2) = \infty$ and lemma 1.13, it follows that for any $\epsilon > 0$, there exist a finite number of points $t_1,\dots,t_n$ and an event A_ϵ', given by equation [1.12], such that

$$\mathbb{P}_1(A_\epsilon') \ge 1-\epsilon, \qquad [1.13]$$

and

$$\mathbb{P}_2(A'_\epsilon) \leq \epsilon. \qquad [1.14]$$

Inequalities [1.13] and [1.14], together with lemma 1.9, imply that $\mathbb{P}_1$ and $\mathbb{P}_2$ are singular.

ii) Let, there exist for any $\epsilon > 0$, both a finite subset $\mathbb{T}' = \{t_1, \dots, t_n\}$ of $\mathbb{T}$ and a number $k \in \{1, \dots, n\}$ such that $\sigma_k < \epsilon$ or $\sigma_k > \frac{1}{\epsilon}$. Let us consider the latter case. For the event

$$A'_\epsilon = \{\omega \in \Omega || Y_k(\omega) - m_k| > \sqrt{\sigma_k}\}$$

belonging to $\mathcal{G} \subset \mathcal{F}$, we obtain the following bounds

$$\mathbb{P}_1(A'_\epsilon) = \mathbb{P}'_1(A'_\epsilon) = \frac{1}{\sqrt{2\pi}\sigma_k} \int_{|y-m_k|>\sqrt{\sigma_k}} e^{-\frac{(y-m_k)^2}{2\sigma_k^2}} dy$$

$$= \frac{1}{\sqrt{2\pi}} \int_{|x|>\sigma_k^{-1/2}} e^{-\frac{x^2}{2}} dx \geq \frac{1}{\sqrt{2\pi}} \int_{|x|>\sqrt{\epsilon}} e^{-\frac{x^2}{2}} dx,$$

and

$$\mathbb{P}_2(A'_\epsilon) = \frac{1}{\sqrt{2\pi}} \int_{|y-m_k|>\sqrt{\sigma_k}} e^{-\frac{x^2}{2}} dx \leq \frac{1}{\sqrt{2\pi}} \int_{|y-m_k|>\epsilon^{-1/2}} e^{-\frac{x^2}{2}} dx.$$

Letting $\epsilon \to 0$, we see that $\mathbb{P}_1(A'_\epsilon) \to 1$ and $\mathbb{P}_2(A'_\epsilon) \to 0$. Again, applying lemma 1.9, we get that $\mathbb{P}_1$ and $\mathbb{P}_2$ are singular. □

THEOREM 1.7.– *Two Gaussian measures on a measurable space* $(\Omega, \mathcal{F})$*, induced by the same Gaussian process, are either equivalent or singular.*

PROOF.– Consider two Gaussian measures $\mathbb{P}_1$ and $\mathbb{P}_2$ on a measurable space $(\Omega, \mathcal{F})$, and let $\mathbf{I}(\mathbb{P}_1, \mathbb{P}_2)$ be the entropy distance between them.

If $\mathbf{I}(\mathbb{P}_1, \mathbb{P}_2) < \infty$, then it follows from definition 1.16 that $\mathbf{H}(\mathbb{P}_1|\mathbb{P}_2) < \infty$ and $\mathbf{H}(\mathbb{P}_2|\mathbb{P}_1) < \infty$. Applying lemma 1.12, we see that $\mathbb{P}_1$ and $\mathbb{P}_2$ are equivalent. Conversely, if $\mathbf{I}(\mathbb{P}_1, \mathbb{P}_2) = \infty$, then it follows from theorem 1.6 that the measures $\mathbb{P}_1$ and $\mathbb{P}_2$ are singular. □

REMARK 1.16.– Theorem 1.7 is very well known as the Feldman–Hajek dichotomy for Gaussian measures, see papers [FEL 58] and [HÁJ 58]. Another proof of this dichotomy, which is obtained with the help of martingale methods, is in [ENG 80].

1.4. Exercises

EXERCISE 1.1.–

1) Consider two random variables X and Y belonging to $L^2(\Omega)$. Prove that $X+Y \in L^2(\Omega)$ while $XY \in L^1(\Omega)$.

2) Establish that $L^2(\Omega)$ is a linear space over $\mathbb{R}$.

EXERCISE 1.2.– *Consider the function $\langle\cdot,\cdot\rangle : L^2(\Omega) \times L^2(\Omega) \to \mathbb{R}$ defined by*

$$\langle f, g\rangle = \mathbb{E}(fg), \text{ where } f, g \in L^2(\Omega).$$

Prove that $\langle\cdot,\cdot\rangle$ is an inner product on $L^2(\Omega)$. Prove that $L^2(\Omega)$ is a Hilbert space w.r.t. the norm generated by this inner product.

EXERCISE 1.3.– *Let a random variable $\xi \sim \mathcal{N}(m,\sigma^2)$. Establish the following formulae for the moments of higher order.*

i) Let $m = 0$, and let $\Gamma(x) = \int_0^\infty t^{x-1}e^{-t}dt, x > 0$ be the Euler's gamma function. Denote $\mathbb{E}_k = \mathbb{E}((\mathcal{N}(0,1)^k), k \in \mathbb{N}$. Then, $E_{2k+1} = 0$, while $E_{2k} = \frac{2^k}{\pi^{1/2}}\Gamma(k+1/2)$. Establish a formula for the moment of any positive (not necessarily integer) order.

ii) Let $m \neq 0, p \in \mathbb{N}$. Then,

$$\mathbb{E}((\mathcal{N}(m,\sigma^2)^{2p}) = \Sigma_{k=0}^{p} C_{2p}^{2k}\sigma^{2k}m^{2p-2k}\mathbb{E}_{2k},$$

and

$$\mathbb{E}((\mathcal{N}(m,\sigma^2)^{2p+1}) = \Sigma_{k=0}^{p} C_{2p+1}^{2k}\sigma^{2k}m^{2p+1-2k}\mathbb{E}_{2k}.$$

EXERCISE 1.4.– *Let $W = \{W_t, t \geq 0\}$ be a Wiener process. Calculate the characteristic function*

$$\mathbb{E}\Big(\exp\Big\{i\sum_{k=1}^{n}\lambda_k W_{t_k}\Big\}\Big),$$

where $n \geq 1$, $\lambda = (\lambda_1,\ldots,\lambda_n) \in \mathbb{R}^n$ and $0 \leq t_1 \leq \cdots \leq t_n$.

EXERCISE 1.5.– *Let $X = \{X_t, t \geq 0\}$ be a Gaussian process with expectation and covariance functions m and C, respectively. Prove that for any $s > 0$, $Y = \{Y_t = X_{t+s} - X_t, t \geq 0\}$ is a Gaussian process and find its expectation and covariance functions.*

EXERCISE 1.6.– *Let $f : \mathbb{R}^+ \to \mathbb{R}$ be a measurable function such that $f \in L^2([0,T],\lambda_1)$, for any $T > 0$.*

1) Define the process $\left\{Y_t = \int_0^t f(s)dW_s, t \geq 0\right\}$ *as a centered, i.e. zero mean, Gaussian process, and let its covariance function equal*

$$C(s,t) := \mathbb{E}(Y_s Y_t) = \int_0^{t\wedge s} f^2(\sigma)d\sigma.$$

Prove that the above definition correctly defines a Gaussian process.

2) Prove that Y *is a Gaussian process with independent increments.*

3) Find $\mathbb{E}\Big(\exp\{i\lambda(Y_t - Y_s)\}\Big)$ *for any* $\lambda \in \mathbb{R}$ *and* $0 \leq s \leq t$.

EXERCISE 1.7.– *Let* $X = \{X_t, t \geq 0\}$ *be a centered Gaussian process with covariance function* $C(s,t)$, $s,t \in \mathbb{R}^+$. *Find* $\mathbb{E}(X_s X_t^2)$, $\mathbb{E}(X_s^3 X_t)$ *and* $\mathbb{E}(X_s^2 X_t^2)$ *for any* $0 \leq s < t$.

EXERCISE 1.8.– *Let* $W^{(1)} = \{W_t^{(1)}, t \geq 0\}$ *and* $W^{(2)} = \{W_t^{(2)}, t \geq 0\}$ *be two independent Wiener processes. Consider the process*

$$W_t := W_t^{(1)} \mathbf{1}_{t\geq 0} + W_{-t}^{(2)} \mathbf{1}_{t\leq 0}. \qquad [1.15]$$

The process defined by equation [1.15] is called a Wiener process on $\mathbb{R}$, *or on the whole real line.*

1) Prove that $\mathbb{E}(W_s W_t) = (|s| \wedge |t|)\mathbf{1}_{st>0}$.

2) Prove that W *is a centered Gaussian process, with covariance function*

$$C(s,t) := \frac{1}{2}(|s| + |t| - |t-s|).$$

3) Prove that W *has independent increments and* $W_t - W_s \sim \mathcal{N}(0, t-s)$ *for any* $s,t \in \mathbb{R},\ s \leq t$.

EXERCISE 1.9.– *Consider a Gaussian process* $X = \{X_t, t \in \mathbb{T}\}$, *and let* $d_X : \mathbb{T} \times \mathbb{T} \to \mathbb{R}_+$ *be its incremental distance, i.e. for any* $(s,t) \in \mathbb{T}^2$,

$$d_X(s,t) = \left(\mathbb{E}(X_t - X_s)^2\right)^{1/2}.$$

Show that $(\mathbb{T}, d_X)$ *is a semi-metric space; see definition A.10 and remark A.5.*

2

Fractional and Sub-fractional Brownian Motions

Throughout this chapter, H stands for some number in the interval $(0, 1)$. This number is called the Hurst index, the Hurst parameter or Hurst exponent, in honor of Harold Edwin Hurst (1880–1978), who was the leading researcher in the field of exact and asymptotic relations demonstrating the phenomena of long memory. In particular, he considered the long-term storage capacity in reservoirs (see [HUR 51]). The Hurst exponent is referred to as the "index of dependence" or "index of long-range dependence".

2.1. Fractional Brownian motion

We will begin studying fractional random processes from the analysis of the following function, which is a candidate for the role of the covariance function. The corresponding assertion can also be found in [NOU 12]. In the following, we denote C, C_H or $C(H)$ different constants, whose value can vary from one statement to another.

LEMMA 2.1.– *The function C defined for any $s, t \in \mathbb{R}$ by means of equality*

$$C(s,t) = \frac{1}{2}\left(|t|^{2H} + |s|^{2H} - |t-s|^{2H}\right), \qquad [2.1]$$

is positive semi-definite.

PROOF.– Let us consider $n \in \mathbb{N}$, $t_1, t_2, \ldots, t_n \in \mathbb{R}$ and $a_1, a_2, \ldots, a_n \in \mathbb{R}$. To get the stated result, we shall prove that

$$\sum_{j,k=1}^{n} a_j a_k C(t_j, t_k) \geq 0.$$

Let us first remark that, if we denote by C_H the finite constant

$$C_H = \int_0^\infty s^{-1-2H}(1-e^{-s^2})ds,$$

then for any real t, we have

$$|t|^{2H} = C_H^{-1}\int_0^\infty s^{-1-2H}(1-e^{-s^2t^2})ds,$$

and consequently, for any $j,k \in \{1,2,\ldots,n\}$,

$$\begin{aligned}
&|t_j|^{2H} + |t_k|^{2H} - |t_j - t_k|^{2H} \\
&= C_H^{-1}\Big(\int_0^\infty s^{-1-2H}(1-e^{-s^2t_j^2})(1-e^{-s^2t_k^2})ds \\
&\quad + \int_0^\infty s^{-1-2H}e^{-s^2(t_j^2+t_k^2)}(e^{2s^2t_jt_k}-1)ds\Big) \\
&= C_H^{-1}\Big(\int_0^\infty s^{-1-2H}(1-e^{-s^2t_j^2})(1-e^{-s^2t_k^2})ds \\
&\quad + \int_0^\infty s^{-1-2H}e^{-s^2(t_j^2+t_k^2)}\sum_{p=1}^\infty \frac{(2s^2t_jt_k)^p}{p!}ds\Big) \\
&= C_H^{-1}\Big(\int_0^\infty s^{-1-2H}(1-e^{-s^2t_j^2})(1-e^{-s^2t_k^2})ds \\
&\quad + \sum_{p=1}^\infty \frac{2^p}{p!}\int_0^\infty s^{-1+2p-2H}t_j^pe^{-s^2t_j^2}t_k^pe^{-s^2t_k^2}ds\Big).
\end{aligned}$$

Therefore,

$$\begin{aligned}
\textstyle\sum_{j,k}^n a_ja_kC(t_j,t_k) &= (2C_H)^{-1}\Big(\textstyle\int_0^\infty s^{-1-2H}\Big(\sum_{k=1}^n a_k(1-e^{-s^2t_k^2}\Big)^2 ds \\
&\quad + \textstyle\sum_{p=1}^\infty \frac{2^p}{p!}\int_0^\infty s^{-1+2p-2H}\Big(\sum_{j=1}^n a_jt_j^pe^{-s^2t_j^2}\Big)^2 ds\Big),
\end{aligned}$$

which is clearly non-negative. □

REMARK 2.1.– The function C defined by [2.1] satisfies $C(s,t) = C(t,s)$ for all $s,t \in \mathbb{R}$. Therefore, according to theorem 1.1, there exists a centered Gaussian process with covariance function $C = C(s,t)$.

DEFINITION 2.1.– *Any centered Gaussian process $B^H = \{B_t^H, t \in \mathbb{R}\}$ with a covariance function defined by equation [2.1] is called a fractional Brownian motion (on $\mathbb{R}$) with the Hurst parameter H.*

REMARK 2.2.– In the particular case where $H = 1/2$, $B^{1/2}$ is the Brownian motion on $\mathbb{R}$ (see exercise 1.8). Therefore, a fractional Brownian motion is an extension of a Brownian motion on $\mathbb{R}$.

In the following proposition, we state some main known properties of the fractional Brownian motion. Their proofs can be found in many sources, see, for example, [NOU 12, NUA 06] and [MIS 08].

PROPOSITION 2.1.– *Let $B^H = \{B_t^H, t \in \mathbb{R}\}$ be a fractional Brownian motion with Hurst parameter H. Then*

1) B^H is a self-similar process with index of self-similarity equal to H (in other words, it is a H-self-similar process). It means that for any $a \in (0, \infty)$, the processes $\{B_{at}^H, t \geq 0\}$ and $\{a^H B_t^H, t \geq 0\}$ have the same probability law.

2) The increments of B^H are stationary, that is, for any $t > 0$, the processes $\{B_{t+s}^H - B_t^H, s \geq 0\}$ and $\{B_s^H, s \geq 0\}$ have the same probability law.

3) The sample paths of B^H are almost surely Hölder continuous with any Hölder exponent $0 < \alpha < H$.

4) For any $H \in (0, 1) \setminus \{1/2\}$, B^H is neither a semi-martingale nor a Markov process.

2.2. Sub-fractional Brownian motion

A sub-fractional Brownian motion (sfBm) is another extension of Brownian motion (Bm), which was investigated by many authors, and a detailed bibliography on the topic is given in the Introduction. SfBm is defined as follows.

DEFINITION 2.2.– *A sub-fractional Brownian motion of Hurst parameter H is a centered Gaussian process $\xi^H = \{\xi_t^H, t \geq 0\}$ with covariance function*

$$\mathrm{Cov}\left(\xi_t^H, \xi_s^H\right) = s^{2H} + t^{2H} - 1/2\left((s+t)^{2H} + |t-s|^{2H}\right). \qquad [2.2]$$

REMARK 2.3.– If $H = 1/2$, then $\xi^{1/2}$ is a Brownian motion on $\mathbb{R}_+$ because in this case,

$$\mathrm{Cov}\left(\xi_t^{1/2}, \xi_s^{1/2}\right) = s + t - 1/2\left((s+t) + |t-s|\right) = s \wedge t.$$

Therefore, a sub-fractional Brownian motion is another extension of a Brownian motion.

REMARK 2.4.– In some applications (such as turbulence phenomena in hydromechanics), fBm is an adequate model for small increments, but it seems to be inadequate for large increments (for more information concerning applications of fBm, see, for example, [MAN 68] and [MON 07]). For this reason, the sub-fractional Brownian motion (sfBm) may be an alternative to fBm in some stochastic models. Moreover, the sfBm arises from occupation time fluctuations of branching particle systems with Poisson initial condition [BOJ 04].

Let us first justify the existence of a sub-fractional Brownian motion.

2.2.1. *Existence of the sfBm*

There are at least two ways to prove the existence of this process. The first one is inspired by [OSS 89], in which the authors proved the existence of the fractional Brownian motion.

To state the existence result, we need the following preliminary lemma.

LEMMA 2.2.– *For fixed $H \in (0,1)$, let*

$$\Gamma_H(t,r) = |t-r|^{H-\frac{1}{2}}\text{sign}(t-r) + 2|r|^{H-\frac{1}{2}}\text{sign}(r) - |t+r|^{H-\frac{1}{2}}\text{sign}(t+r)$$

for $t,r \in \mathbb{R}$. Then, for any $(s,t) \in \mathbb{R}^2$, $\Gamma_H(t,\cdot) \in L^2(\mathbb{R},\lambda_1)$ and

$$\langle \Gamma_H(t,\cdot), \Gamma_H(s,\cdot)\rangle_{L^2(\mathbb{R},\lambda_1)} = 2C_H\Big(s^{2H} + t^{2H} - \frac{1}{2}(|s+t|^{2H} + |s-t|^{2H})\Big),$$

where C_H is a positive constant and $\langle \cdot,\cdot\rangle_{L^2(\mathbb{R},\lambda_1)}$ denotes the inner product in the space $L^2(\mathbb{R},\lambda_1)$ defined by $\langle f,g\rangle_{L^2(\mathbb{R},\lambda_1)} = \int_{\mathbb{R}} f(r)g(r)dr$.

PROOF.– Note that $\Gamma_H(0,r) = 0$ and $\big(\Gamma_H(t,r)\big)^2 = |t|^{2H-1}\left(\Gamma_H\left(1,\frac{r}{t}\right)\right)^2$ for any $r \in \mathbb{R}$ and $t \in \mathbb{R}\setminus\{0\}$. We also note that the function

$$u \longmapsto \Gamma_H^2(1,u) = \Big(|1-u|^{H-\frac{1}{2}}\text{sign}(1-u) + 2|u|^{H-\frac{1}{2}}\text{sign}(u) \\ -|1+u|^{H-\frac{1}{2}}\text{sign}(1+u)\Big)^2$$

is locally integrable on $\mathbb{R}\setminus\{-1,0,1\}$. Moreover, since $H \in (0,1)$, the functions $u \longmapsto |1-u|^{H-\frac{1}{2}}\text{sign}(1-u)$, $u \longmapsto |u|^{H-\frac{1}{2}}\text{sign}(u)$ and $u \longmapsto |1+u|^{H-\frac{1}{2}}\text{sign}$ $(1+u)$ are locally square integrable in neighborhoods of $u=-1, u=0$ and $u=1$. Therefore, $\int_{-2}^{2}\Gamma_H^2(1,u)du < \infty$. In addition, note that $\Gamma_H(1,-u) = -\Gamma_H(1,u)$.

Therefore, we have $\Gamma_H^2(1,-u) = \Gamma_H^2(1,u)$. Taking into account this equality and the fact that $2H-3<-1$, we obtain the following bounds:

$$\int_{\mathbb{R}\setminus[-2,2]} \Gamma_H^2(1,u)du = 2\int_2^{+\infty} \Gamma_H^2(1,u)du$$

$$= 2\int_2^{+\infty} \left(\left(u^{H-\frac{1}{2}} - (u-1)^{H-\frac{1}{2}}\right) - \left((u+1)^{H-\frac{1}{2}} - u^{H-\frac{1}{2}}\right)\right)^2 du$$

$$\leq 4\int_2^{+\infty} \left(u^{H-\frac{1}{2}} - (u-1)^{H-\frac{1}{2}}\right)^2 du + 4\int_2^{+\infty} \left((u+1)^{H-\frac{1}{2}} - u^{H-\frac{1}{2}}\right)^2 du$$

$$\leq 4\int_2^{+\infty} \left(\int_{u-1}^{u} \left(H-\frac{1}{2}\right) y^{H-\frac{3}{2}}dy\right)^2 du$$

$$+ 4\int_2^{+\infty} \left(\int_{u}^{u+1} \left(H-\frac{1}{2}\right) y^{H-\frac{3}{2}}dy\right)^2 du$$

$$\leq 4\left(H-\frac{1}{2}\right)^2 \left(\int_2^{+\infty}\int_{u-1}^{u} y^{2H-3}dydu + \int_2^{+\infty}\int_{u}^{u+1} y^{2H-3}dydu\right)$$

$$\leq 4\left(H-\frac{1}{2}\right)^2 \left(\int_2^{+\infty}(u-1)^{2H-3}du + \int_2^{+\infty} u^{2H-3}du\right) < \infty.$$

Therefore, $\Gamma_H(t,\cdot) \in L^2(\mathbb{R},\lambda_1)$ for any $t\in\mathbb{R}$.

Now, in order to evaluate $\langle \Gamma_H(t,\cdot), \Gamma_H(s,\cdot)\rangle_{L^2(\mathbb{R},\lambda_1)}$, we first note that

$$\Gamma_H(t,\cdot) := \gamma_H(t,\cdot) + \gamma_H(-t,\cdot),\ t\in\mathbb{R}, \qquad [2.3]$$

where $\gamma_H(t,r) = |t-r|^{H-\frac{1}{2}}\text{sign}(t-r) + |r|^{H-\frac{1}{2}}\text{sign}(r)$, for any $(t,r)\in\mathbb{R}^2$.

It is easy to see that $\gamma_H(0,r) = 0$ and $\left(\gamma_H(t,r)\right)^2 = |t|^{2H-1}\left(\gamma_H\left(1,\frac{r}{t}\right)\right)^2$ for any $r \in \mathbb{R}$ and $t\in\mathbb{R}\setminus\{0\}$. Similarly to Γ_H, it is easy to check that $\gamma_H(t,\cdot)\in L^2(\mathbb{R},d\lambda_1)$ for any t, and

$$\|\gamma_H(t,\cdot)\|^2_{L^2(\mathbb{R},\lambda_1)} = \int_{\mathbb{R}} |t|^{2H-1}\gamma_H^2\left(1,\frac{r}{t}\right)dr$$

$$= |t|^{2H}\int_{\mathbb{R}}\gamma_H^2(1,u)du = |t|^{2H}C_H, \qquad [2.4]$$

with $C_H = \|\gamma_H^2(1,\cdot)\|^2_{L^2(\mathbb{R},\lambda_1)} > 0$.

From the linearity of $L^2(\mathbb{R}, \lambda_1)$, we deduce that $\gamma_H(t,\cdot) - \gamma_H(s,\cdot) \in L^2(\mathbb{R}, \lambda_1)$ and

$$\begin{aligned}
&\|\gamma_H(t,\cdot) - \gamma_H(s,\cdot)\|^2_{L^2(\mathbb{R},\lambda_1)} \\
&= \int_{\mathbb{R}} \left(|t-r|^{H-\frac{1}{2}}\operatorname{sign}(t-r) - |s-r|^{H-\frac{1}{2}}\operatorname{sign}(s-r)\right)^2 dr \\
&= \int_{\mathbb{R}} \left(|t-s+u|^{H-\frac{1}{2}}\operatorname{sign}(t-s+u) - |u|^{H-\frac{1}{2}}\operatorname{sign}(u)\right)^2 du \\
&= \int_{\mathbb{R}} \left(|t-s-v|^{H-\frac{1}{2}}\operatorname{sign}(t-s-v) + |v|^{H-\frac{1}{2}}\operatorname{sign}(v)\right)^2 dv \\
&= \|\gamma_H(t-s,\cdot)\|^2_{L^2(\mathbb{R},\lambda_1)} = C_H|t-s|^{2H}.
\end{aligned} \qquad [2.5]$$

So, it follows from the bilinear property of the inner product, together with the equalities [2.4] and [2.5], that

$$\begin{aligned}
\langle \gamma_H(t,\cdot), \gamma_H(s,\cdot)\rangle_{L^2(\mathbb{R},\lambda_1)} &= \frac{1}{2}\Bigg(\|\gamma_H(t,\cdot)\|^2_{L^2(\mathbb{R},\lambda_1)} + \|\gamma_H(s,\cdot)\|^2_{L^2(\mathbb{R},\lambda_1)} \\
-\|\gamma_H(t,\cdot) - \gamma_H(s,\cdot)\|^2_{L^2(\mathbb{R},\lambda_1)}\Bigg) &= \frac{C_H}{2}\left(|t|^{2H} + |s|^{2H} - |t-s|^{2H}\right).
\end{aligned} \qquad [2.6]$$

Now, by using equations [2.3] and [2.6], we get:

$$\begin{aligned}
&\langle \Gamma_H(t,\cdot), \Gamma_H(s,\cdot)\rangle_{L^2(\mathbb{R},\lambda_1)} \\
&\quad = \langle \gamma_H(t,\cdot) + \gamma_H(-t,\cdot), \gamma_H(s,\cdot) + \gamma_H(-s,\cdot)\rangle_{L^2(\mathbb{R},\lambda_1)} \\
&\quad = \langle \gamma_H(t,\cdot), \Gamma_H(s,\cdot)\rangle_{L^2(\mathbb{R},\lambda_1)} + \langle \gamma_H(t,\cdot), \gamma_H(-s,\cdot)\rangle_{L^2(\mathbb{R},\lambda_1)} \\
&\quad + \langle \gamma_H(-t,\cdot), \Gamma_H(s,\cdot)\rangle + \langle \gamma_H(-t,\cdot), \gamma_H(-s,\cdot)\rangle_{L^2(\mathbb{R},\lambda_1)} \\
&\quad = \frac{C_H}{2}\Bigg(|t|^{2H} + |s|^{2H} - |t-s|^{2H} + |t|^{2H} + |-s|^{2H} - |t-(-s)|^{2H} \\
&\quad + |-t|^{2H} + |s|^{2H} - |-t-s|^{2H} + |-t|^{2H} + |-s|^{2H} \\
&\quad - |-t-(-s)|^{2H}\Bigg) = 2C_H(t^{2H} + s^{2H} - \frac{1}{2}\left(|t-s|^{2H} + |t+s|^{2H}\right),
\end{aligned}$$

and the proof follows. □

Now, we can state the result justifying the existence of sfBm.

THEOREM 2.1.– *The function G defined for any $s, t \in \mathbb{R}_+$ by the equality*

$$G(s,t) = s^{2H} + t^{2H} - 1/2\Big((s+t)^{2H} + |t-s|^{2H}\Big), \qquad [2.7]$$

is positive semi-definite.

PROOF.– Let us consider $n \in \mathbb{N}, t_1, t_2, \ldots, t_n \in \mathbb{R}$ and $a_1, a_2, \ldots, a_n \in \mathbb{R}$. It follows from lemma 2.2 that

$$\sum_{j,k=1}^{n} a_j a_k G(t_j, t_k) = \sum_{j,k=1}^{n} a_j a_k \frac{1}{2C_H} \langle \Gamma_H(t_j, \cdot), \Gamma_H(t_k, \cdot) \rangle_{L^2(\mathbb{R},\lambda_1)}$$
$$= \frac{1}{2C_H} \left\| \sum_{j=1}^{n} a_j \Gamma_H(t_j, \cdot) \right\|^2_{L^2(\mathbb{R},\lambda_1)} \geq 0.$$

□

REMARK 2.5.– Function G defined by equation [2.7] satisfies $G(s,t) = G(t,s)$ for all $s, t \in \mathbb{R}$. Therefore, it follows from theorem 1.1 that there exists a centered Gaussian process with covariance function G.

We now present a second method to establish the existence of the sfBm. Consider the process $\xi^H = \{\xi_t^H, t \geq 0\}$ defined by

$$\xi_t^H = \frac{B_t^H + B_{-t}^H}{\sqrt{2}}, \; t \geq 0, \qquad [2.8]$$

where $H \in (0,1)$, and $\{B_t^H, \; t \in \mathbb{R}\}$ is a fractional Brownian motion (fBm) on $\mathbb{R}$.

LEMMA 2.3.– *The process ξ^H defined by equation [2.8] is a sub-fractional Brownian motion.*

PROOF.– The fBm B^H is Gaussian and centered. Thus, it follows from equation [2.8] that ξ^H is also a centered Gaussian process. Moreover, we can conclude from equation [2.8] that the covariance function of B^H is defined by equation [2.1], whence, by making a simple calculation, we obtain that the covariance functions of ξ^H satisfy equation [2.2]. □

LEMMA 2.4.– *Let the probability space $(\Omega, \mathcal{F}, \mathbb{P})$ be rich enough to support any of stochastic processes introduced below, and let $\xi^H = \{\xi_t^H, t \geq 0\}$ be a sfBm. Then, there exists a fractional Brownian motion $B^H = \{B_t^H, t \in \mathbb{R}\}$ on $\mathbb{R}$ such that ξ^H admits representation [2.8].*

PROOF.– On the same probability space, consider an independent centered Gaussian process $Z^H = \{Z_t^H, t \geq 0\}$ with covariance function

$$\mathrm{Cov}\left(Z_t^H, Z_s^H\right) := \frac{1}{2}\left((t+s)^{2H} - |t-s|^{2H}\right). \qquad [2.9]$$

Note that the right-hand side of equation [2.9] indeed represents a covariance function because for any fractional Brownian motion $\tilde{B}^H$ on $\mathbb{R}$, and $s,t \in \mathbb{R}$,

$$(t+s)^{2H} - |t-s|^{2H} = 2\mathbb{E}\left((\tilde{B}_t^H - \tilde{B}_{-t}^H)(\tilde{B}_s^H - \tilde{B}_{-s}^H)\right).$$

Now, create a centered Gaussian process

$$B_t^H = \begin{cases} \frac{1}{\sqrt{2}}(\xi_t^H + Z_t^H) \text{ if } t \geq 0, \\ \frac{1}{\sqrt{2}}(\xi_{-t}^H - Z_{-t}^H) \text{ if } t \leq 0. \end{cases} \qquad [2.10]$$

It is very easy to check that B^H is a fractional Brownian motion on $\mathbb{R}$. For example, consider $\mathbb{E}(B_{-s}^H B_t^H)$ for $s,t \geq 0$:

$$\begin{aligned} \mathbb{E}(B_{-s}^H B_t^H) &= \tfrac{1}{2}\mathbb{E}\left((\xi_{-t}^H - Z_{-t}^H)(\xi_t^H + Z_t^H)\right) \\ &= \tfrac{1}{2}\left(s^{2H} + t^{2H} - \tfrac{1}{2}(t+s)^{2H} - \tfrac{1}{2}|t-s|^{2H} - \tfrac{1}{2}(t+s)^{2H} + \tfrac{1}{2}|t-s|^{2H}\right) \\ &= \tfrac{1}{2}\left(|s|^{2H} + |t|^{2H} - |t-(-s)|^{2H}\right). \end{aligned}$$

Therefore, B^H is a fractional Brownian motion on $\mathbb{R}$, and from formula [2.10], we immediately get that

$$\xi_t^H = \frac{1}{\sqrt{2}}(B_t^H + B_{-t}^H), t \geq 0.$$

□

2.2.2. *Main properties of the sfBm*

Recall again that a sfBm $\xi^H = \{\xi_t^H, t \geq 0\}$ is a zero-mean Gaussian process with covariance function

$$\text{Cov}\left(\xi_t^H, \xi_s^H\right) = s^{2H} + t^{2H} - 1/2\left((s+t)^{2H} + |t-s|^{2H}\right).$$

It is very easy to produce the following calculations: for any $t \geq 0$

$$\mathbb{E}(\xi_t^H)^2 = (2 - 2^{2H-1})t^{2H},$$

and for any $0 \leq s \leq t$

$$\mathbb{E}\left(\xi_t^H - \xi_s^H\right)^2 = -2^{2H-1}(t^{2H} + s^{2H}) + (t+s)^{2H} + (t-s)^{2H}. \qquad [2.11]$$

Now, we can establish the following interesting result.

LEMMA 2.5.– *There exist two positive constants C_1 and C_2 depending on H such that for all $0 \leq s \leq t$*

$$C_1(t-s)^{2H} \leq \mathbb{E}(\xi_t^H - \xi_s^H)^2 \leq C_2(t-s)^{2H}.$$

PROOF.–

i) Let $0 < H < 1/2$. Then,

$$(t+s)^{2H} - (t-s)^{2H} \leq (2s)^{2H} \leq 2^{2H-1}\left(t^{2H} + s^{2H}\right),$$

whence

$$\mathbb{E}\left(\xi_t^H - \xi_s^H\right)^2 \leq 2(t-s)^{2H}.$$

Therefore, for $0 < H < 1/2$, we have that $C_2 = 2$. Furthermore, in this case, the function $f(t) = t^{2H}$ is concave, therefore

$$\left(\frac{t+s}{2}\right)^{2H} \geq \frac{t^{2H} + s^{2H}}{2},$$

and consequently,

$$\mathbb{E}\left(\xi_t^H - \xi_s^H\right)^2 \geq (t-s)^{2H},$$

so $C_1 = 1$.

ii) Let $1/2 < H < 1$. Then, the function $f(t) = t^{2H}$ is convex, therefore

$$\left(\frac{t+s}{2}\right)^{2H} \leq \frac{t^{2H} + s^{2H}}{2},$$

and consequently,

$$\mathbb{E}\left(\xi_t^H - \xi_s^H\right)^2 \leq (t-s)^{2H},$$

so, in this case, $C_2 = 1$. The lower bound is a little bit more involved. First, remark that it holds for $t = 0$. Furthermore, let $t > 0$. Then, we can divide the left- and right-hand sides of the desirable inequality by t^{2H} and get that our goal is to establish the lower bound

$$-2^{2H-1}(1 + z^{2H}) + (1 + z)^{2H} + (1 - z)^{2H} \geq C_1(1 - z)^{2H}, 0 \leq z \leq 1.$$

Note that the left- and right-hand sides of the inequality above are equal for $z = 0$ if we put $C_1 = 2 - 2^{2H-1}$. With this constant, the derivatives of the left- and right-hand sides equal, up to the multiplier $2H$,

$$-2^{2H-1}z^{2H-1} + (1 + z)^{2H-1} - (1 - z)^{2H-1}, \text{ and } -(2 - 2^{2H-1})(1 - z)^{2H-1},$$

correspondingly. Thus, it is sufficient to prove that

$$-2^{2H-1}z^{2H-1} + (1 + z)^{2H-1} - (1 - z)^{2H-1} \geq -(2 - 2^{2H-1})(1 - z)^{2H-1},$$

or, as is equivalent,

$$(1 + z)^{2H-1} \geq (2^{2H-1} - 1)(1 - z)^{2H-1} + 2^{2H-1}z^{2H-1}.$$

However, because $0 < 2H - 1 < 1$, we have that the function $g(t) = t^{2H-1}$ is concave, therefore,

$$\left(\frac{1 + z}{2}\right)^{2H-1} \geq \frac{1 + z^{2H-1}}{2},$$

whence

$$\begin{aligned}&(1 + z)^{2H-1} \geq 2^{2H-1}(1 + z^{2H-1})\\&= 2^{2H-1} + 2^{2H-1}z^{2H-1} \geq (2^{2H-1} - 1)(1 - z)^{2H-1} + 2^{2H-1}z^{2H-1},\end{aligned}$$

therefore, indeed, constant $C_1 = 2 - 2^{2H-1}$, and the lemma is proved. □

COROLLARY 2.1.– *Lemma 2.5 states that an sfBm belongs to quasi-helix*

$$\mathbb{QH}(H, C_1, C_2, [0, T]),$$

see definition 1.11. Then, obviously, the sample paths of ξ^H *are almost surely Hölder continuous with any Hölder exponent* $0 < \alpha < H$, *see example 1.2. Lemma 2.5 also implies that an sfBm is a non-degenerate Gaussian process for which the incremental distance* $d_{\xi^H}(s,t) = \left(\mathbb{E}(\xi_t^H - \xi_s^H)^2\right)^{1/2}$ *is a metric. Moreover, we can apply theorem 1.5 to a sfBm and state the following result.*

PROPOSITION 2.2.–

i) For any $H \in (0, 1/2)$

$$\frac{1}{5\sqrt{H}} \le \mathbb{E} \max_{0 \le t \le 1} \xi_t^H \le \frac{32.6}{\sqrt{H}}.$$

i) For any $H \in (1/2, 1)$

$$\frac{2 - 2^{2H-1}}{5\sqrt{H}} \le \mathbb{E} \max_{0 \le t \le 1} \xi_t^H \le \frac{16.3}{\sqrt{H}}.$$

REMARK 2.6.– Of course, a similar result also holds for a fractional Brownian motion, but the constants will be different: for any $H \in (0, 1)$

$$\frac{1}{5\sqrt{H}} \le \mathbb{E} \max_{0 \le t \le 1} B_t^H \le \frac{16.3}{\sqrt{H}}.$$

It means, in particular, that

$$\mathbb{E} \max_{0 \le t \le 1} \xi_t^H \to \infty \text{ and } \mathbb{E} \max_{0 \le t \le 1} B_t^H \to \infty$$

as $H \to 0$.

The following additional properties of sfBm were established in [BOJ 04] and [TUD 07].

LEMMA 2.6.– *A sfBm* $\{\xi_t^H, t \in \mathbb{R}_+\}$ *satisfies the following properties:*

i) ξ^H *is a self-similar process of the index of self-similarity that equals* H.

ii) For any $H \in (0, 1) \setminus \{1/2\}$, ξ^H *is neither a semi-martingale nor a Markov process.*

iii) For any $0 < \epsilon < H$, *there exists a non-negative random variable* $G_{H,\epsilon,T}$ *such that* $E(G_{H,\epsilon,T}^p) < \infty$ *for any* $p \ge 1$, *and*

$$|\xi_t^H - \xi_s^H| \le G_{H,\epsilon,T} |t - s|^{H-\epsilon},$$

a.s. for all $s, t \in [0, T]$.

Let us remark that the properties stated in lemma 2.6 are also satisfied by fractional Brownian motion. In particular, property (iii) for a fBm was established in [NUA 02].

Now consider a property that marks a main difference between the two processes, fBm and sfBm.

LEMMA 2.7.– *The sfBm does not have stationary increments unless* $H = \frac{1}{2}$.

PROOF.– Let $H \neq \frac{1}{2}$ (otherwise the situation is evident since we have a Wiener process). The statement can be immediately deduced from equation [2.11] because the latter equality implies that, generally speaking,

$$\mathbb{E}\Big(\xi_{t+h}^H - \xi_{s+h}^H\Big)^2 \neq \mathbb{E}\Big(\xi_t^H - \xi_s^H\Big)^2.$$

Indeed, let $s = 0$ and $t = h > 0$. Then

$$\mathbb{E}(\xi_t^H - \xi_s^H)^2 = \mathbb{E}(\xi_h^H)^2 = (2 - 2^{2H-1})h^{2H},$$

and

$$\begin{aligned}\mathbb{E}(\xi_{t+h}^H - \xi_{s+h}^H)^2) &= \mathbb{E}(\xi_{2h}^H - \xi_h^H)^2 \\ &= (3h)^{2H} + h^{2H} - 2^{2H-1}((2h)^{2H} + h^{2H}) \\ &= (3^{2H} + 1 - 2^{4H-1} - 2^{2H-1})h^{2H}.\end{aligned}$$

Now, for $H \in [0, 1]$, evaluate the difference

$$a(H) := 3^{2H} + 1 - 2^{4H-1} - 2^{2H-1} - (2 - 2^{2H-1}) = 3^{2H} - 2^{4H-1} - 1.$$

Obviously, $a(0) = -\frac{1}{2} < 0$ and $a(\frac{1}{2}) = a(1) = 0$. In order to differentiate, we slightly rewrite $a(H)$ replacing $2H$ with x and consider the function

$$\tilde{a}(x) = 3^x - \frac{4^x}{2} - 1; \; x \in [0, 2].$$

Then

$$\tilde{a}'(x) = 3^x \log x - \frac{4^x}{2} \log 4 = 4^x \log 3 \left(\left(\frac{3}{4}\right)^x - \frac{\log 2}{\log 3}\right).$$

The function $x \longmapsto \left(\frac{3}{4}\right)^x - \frac{\log 2}{\log 3}$ is decreasing from $1 - \frac{\log 2}{\log 3}$ to $\frac{9}{16} - \frac{\log 2}{\log 3} = 0.5625 - 0.6309 < 0$. Therefore, there is one root $x_0 \in (0, 2)$ of the equation $\tilde{a}'(x) = 0$. Consequently, $x \longmapsto \tilde{a}(x)$ is increasing on the interval $[0, x_0]$ and decreasing on $[x_0, 2]$. This means that the trajectory of $a(H)$ is as in Figure 2.1.

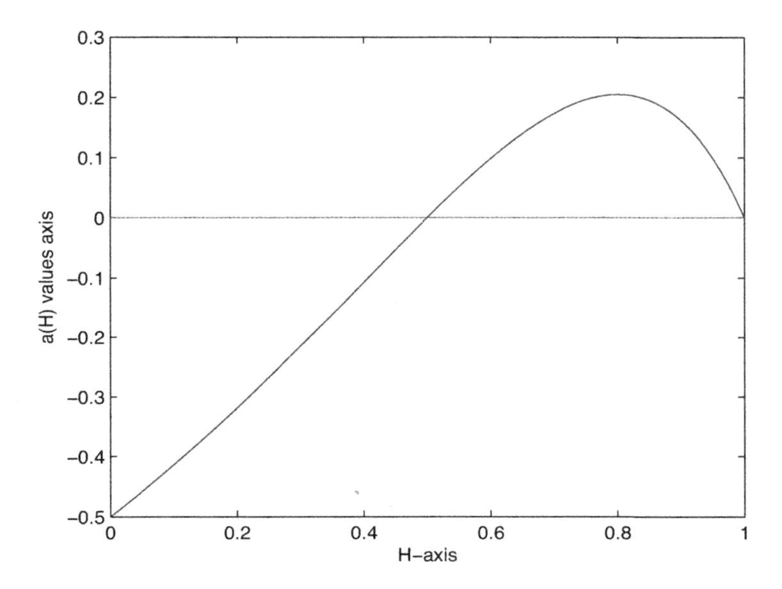

Figure 2.1. *Graph of a(H)*

Thus, $a(H)$ is negative for $H \in [0, \frac{1}{2})$ and strictly positive for $H \in \left(\frac{1}{2}, 1\right]$. This means that

$$\mathbb{E}(\xi_{2h}^H - \xi_h^H)^2 < \mathbb{E}(\xi_h^H)^2 \text{ for } H \in \left(0, \frac{1}{2}\right) \text{ and}$$

$$\mathbb{E}(\xi_{2h}^H - \xi_h^H)^2 > \mathbb{E}(\xi_h^H)^2 \text{ for } H \in \left(\frac{1}{2}, 1\right). \qquad \square$$

2.3. Long- and short-range dependence of fBm and sfBm

DEFINITION 2.3.– *A centered Gaussian sequence $\eta = \{\eta_n, n \geq 1\}$ is called long-range dependent or short-range dependent depending on*

$$\sum_{n=1}^{\infty} |\mathbb{E}\eta_1\eta_n| = \infty \text{ or } \sum_{n=1}^{\infty} |\mathbb{E}\eta_1\eta_n| < \infty,$$

respectively.

More often, the notion of long- and short-range dependence is applied to Gaussian stationary sequences. However, we will apply it to sfBm as well. A fractional Brownian motion B^H will be considered as a process with long (short)-range dependence if $\sum_{n=1}^{\infty} |\mathbb{E}B_1^H(B_{n+1}^H - B_n^H)| = \infty$ $\left(\sum_{n=1}^{\infty} |\mathbb{E}B_1^H(B_{n+1}^H - B_n^H)| < \infty\right)$. The same definition will be considered for sfBm.

LEMMA 2.8.– *A fractional Brownian motion is long (short)-range dependent if and only if* $H > 1/2$ *(*$H < 1/2$*). Any sfBm is a short-range-dependent stochastic process.*

PROOF.– Applying Taylor expansion to the power function, we get, for a fractional Brownian motion:

$$\begin{aligned}\mathbb{E}B_1^H(B_{n+1}^H - B_n^H) &= \frac{1}{2}\left(1 + (n+1)^{2H} - n^{2H} - 1 - n^{2H} + (n-1)^{2H}\right)\\ &= \frac{1}{2}n^{2H}\left(\left(1+\frac{1}{n}\right)^{2H} + \left(1-\frac{1}{n}\right)^{2H} - 2\right)\\ &= \frac{1}{2}n^{2H}\left(2H(2H-1)n^{-2} + o(n^{-2})\right)\\ &= H(2H-1)n^{2H-2} + o(n^{2H-2}),\end{aligned}$$

whence the statement of lemma 2.8 concerning a fractional Brownian motion follows immediately. Similarly, applying Taylor expansion to the power function, we get, for a sub-fractional Brownian motion:

$$\begin{aligned}\mathbb{E}\xi_1^H(\xi_{n+1}^H - \xi_n^H) &= 1 + (n+1)^{2H} - \frac{1}{2}\left((n+2)^{2H} + n^{2H}\right)\\ &\quad - 1 - n^{2H} + \frac{1}{2}\left((n+1)^{2H} + (n-1)^{2H}\right)\\ &= \frac{3}{2}\left(n+1\right)^{2H} - \frac{3}{2}n^{2H} + \frac{1}{2}\left(n-1\right)^{2H}\\ &\quad - \frac{1}{2}\left(n+2\right)^{2H} = \frac{1}{2}n^{2H}\left(3\left(1+\frac{1}{n}\right)^{2H}\right.\\ &\quad \left. -3 + \left(1-\frac{1}{n}\right)^{2H} - \left(1+\frac{2}{n}\right)^{2H}\right)\\ &= 2H(1-H)(2H-1)n^{2H-3} + o(n^{2H-3}),\end{aligned}$$

whence the statement of lemma 2.8 concerning a sub-fractional Brownian motion follows. □

2.4. "Moving average" representation of fBm and sfBm

The proof of the following result can be found, for example, in [SAM 94], proposition 7.2.6. For the calculation of the constant, see [MIS 08].

PROPOSITION 2.3.– *Let $0 < H < 1$ and $B^H = \{B_t^H, t \in \mathbb{R}\}$ be a fBm on $\mathbb{R}$. Then, B^H has the following integral representation:*

$$B_t^H = C_1(H) \int_{\mathbb{R}} \left((t-s)_+^{H-\frac{1}{2}} - (-s)_+^{H-\frac{1}{2}} \right) dW_s,$$

where W is a Wiener process on $\mathbb{R}$, and

$$C_1(H) = \left(\int_0^\infty \left((1+x)^{H-\frac{1}{2}} - x^{H-\frac{1}{2}} \right)^2 dx + \frac{1}{2H} \right)^{-1/2}$$
$$= \frac{(2H \sin(\pi H)\Gamma(2H))^{1/2}}{\Gamma(H+1/2)}.$$

As a consequence of proposition 2.3, we get the following statement.

COROLLARY 2.2.– *Let $0 < H < 1$ and $\xi^H = \{\xi_t^H, t \in \mathbb{R}_+\}$ be a sub-fractional Brownian motion with Hurst parameter H. Then, ξ^H has the following integral representation: for any $t \geq 0$*

$$\xi_t^H = \frac{C_1(H)}{\sqrt{2}} \int_{\mathbb{R}} \left((t-s)_+^{H-\frac{1}{2}} + (t+s)_-^{H-\frac{1}{2}} - 2(-s)_+^{H-\frac{1}{2}} \right) dW_s.$$

PROOF.– It follows from lemma 2.4 that there exists an fBm B^H such that

$$\xi_t^H = \frac{B_t^H + B_{-t}^H}{\sqrt{2}}, \text{ for any } t \geq 0.$$

Then, it follows from the latter representation and from proposition 2.3 that

$$\xi_t^H = \frac{C_1(H)}{\sqrt{2}} \left(\int_{\mathbb{R}} \left((t-s)_+^{H-\frac{1}{2}} - (-s)_+^{H-\frac{1}{2}} \right) dW_s \right.$$
$$\left. + \int_{\mathbb{R}} \left((-t-s)_+^{H-\frac{1}{2}} - (-s)_+^{H-\frac{1}{2}} \right) dW_s \right)$$
$$= \frac{C_1(H)}{\sqrt{2}} \int_{\mathbb{R}} \left((t-s)_+^{H-\frac{1}{2}} + (t+s)_-^{H-\frac{1}{2}} - 2(-s)_+^{H-\frac{1}{2}} \right) dW_s,$$

and the proof follows. □

2.5. Spectral representation of fBm and sfBm

Let us start with a simple and even non-random example to explain the meaning of the spectral decomposition (spectral representation, harmonizable representation) of a stochastic process. Thus, recall that the Fourier transform of a real signal $f = f(t)$ is given by the complex exponential

$$F(x) = \int_{\mathbb{R}} e^{-ixt} f(t)dt,$$

under assumption $f \in L^1(\mathbb{R}, \lambda_1)$. In turn, under assumption $F \in L^1(\mathbb{R}, \lambda_1)$, the signal f can be reconstructed from F by the inverse Fourier transform as follows:

$$f(t) = \frac{1}{2\pi} \int_{\mathbb{R}} e^{ixt} F(x)dx, \qquad [2.12]$$

The existence of this inverse Fourier transform means that we can represent a signal f as a superposition of continuum of complex sinusoids. The Fourier transform $F(x)$ is the strength of the complex exponents of frequency x presented in the signal. The function F is also called the spectrum of f, and representation [2.12] is called a representation of a signal in a frequency domain. Now, how to get the frequency-domain representation of a stochastic process? To begin with, we consider and compare the spectral representations of the covariance increment function of a fBm and of a sfBm.

EXAMPLE 2.1.– *Consider a fractional Brownian motion* $B^H = \{B_t^H, t \in \mathbb{R}\}$ *and a fractional Gaussian noise* $\zeta^H = \{\zeta_j^H = B_{j+1}^H - B_j^H, j \in \mathbb{Z}\}$. *This is a stationary sequence, whose covariance function equals*

$$r_j := \mathbb{E}\zeta_1^H \zeta_j^H = 1/2(|j+1|^H + |j-1|^H - 2|j|^H)$$

and has a spectral density $h(\lambda)$. *In other words, there exists a spectral representation*

$$r_j = \int_{-\pi}^{\pi} e^{ij\lambda} h(\lambda)d\lambda,$$

where

$$h(\lambda) = C\sin^2\left(\frac{\lambda}{2}\right)\left(|\lambda|^{-2H-1} + \sum_{k\in\mathbb{Z}, k\neq 0} |\lambda + 2\pi k|^{-2H-1}\right).$$

For details, see [SAM 94]. This spectral density explodes at zero for $H \in (0, 1/2)$ *and is smooth at zero for* $H \in (1/2, 1)$, *corresponding to the short- or long-range dependence of the respective fractional Brownian motion, respectively. Now, consider a sfBm* ξ^H *and create a sub-fractional Gaussian noise* $\vartheta^H = \{\vartheta_i^H = \xi_{i+1}^H - \xi_i^H, i \in \mathbb{Z}\}$. *Despite its non-stationarity, we can, of course, consider the similar function*

$$\begin{aligned} R_j &:= \mathbb{E}\vartheta_1^H \vartheta_j^H = 1/2(\mathbb{E}(B_1^H + B_{-1}^H)(B_{j+1}^H + B_{-j-1}^H - B_j^H - B_{-j}^H)) \\ &= \mathbb{E}(B_1^H + B_{-1}^H)(B_{j+1}^H - B_j^H) = r_j - r_{j+1}, j \in \mathbb{Z}. \end{aligned}$$

For this function, we have a "spectral representation"

$$R_j = \int_{-\pi}^{\pi} e^{ij\lambda}(1 - e^{i\lambda})h(\lambda)d\lambda,$$

and the "spectral density" $(1 - e^{i\lambda})h(\lambda)$ *has the following asymptotic behavior at zero:*

$$(1 - e^{i\lambda})h(\lambda) \sim C|\lambda|^3|\lambda|^{-2H-1} \sim C|\lambda|^{2-2H}.$$

So, $(1 - e^{i\lambda})h(\lambda)$ *is smooth at zero for any* $H \in (0, 1)$, *which corresponds to the fact that sfBm has a short-range dependence for any* $H \in (0, 1)$.

Now, to switch to spectral representations of a fBm and a sfBm, we establish the following characterization lemma.

LEMMA 2.9.– *Consider* $H \in (0, 1)$ *and a Gaussian process* $X = \{X_t, t \in \mathbb{R}\}$ *satisfying* $\mathbb{E}(X^2(1)) = 1$. *Then,* X *is a fractional Brownian motion with Hurst parameter* H *if and only if* X *is* H*-self-similar and has stationary increments.*

PROOF.– We know from proposition 2.1 that any fractional Brownian motion with Hurst parameter H is a H-self-similar Gaussian process with stationary increments. Inversely, let X be a Gaussian process, H-self-similar and with stationary increments. Then, on the one hand, it follows from the stationarity of the increments that for any $t \geq s$

$$\begin{aligned} \mathbb{E}(X_s X_t) &= \frac{1}{2}\left(\mathbb{E}(X_t^2) + \mathbb{E}(X_s^2) - \mathbb{E}(X_t - X_s)^2\right) \\ &= \frac{1}{2}\left(\mathbb{E}(X_t^2) + \mathbb{E}(X_s^2) - \mathbb{E}(X_{t-s} - X_0)^2\right). \end{aligned}$$

This equality, together with the self-similarity, implies that

$$\begin{aligned} \mathbb{E}(X_s X_t) &= \frac{1}{2}\left(|t|^{2H} + |s|^{2H} - |t - s|^{2H}\right)\mathbb{E}(X_1^2) \\ &= \frac{1}{2}\left(|t|^{2H} + |s|^{2H} - |t - s|^{2H}\right). \end{aligned}$$

On the other hand, it follows from proposition 1.3 that $X_0 = 0$ a.s. and X_t has the same distribution as $t^H X_1$. Therefore,

$$\mathbb{E}(X_1) = \mathbb{E}(X_2 - X_1) = 2^H\mathbb{E}(X_1) - \mathbb{E}(X_1) = (2^H - 1)\mathbb{E}(X_1),$$

which implies that $\mathbb{E}(X_1) = 0$.

Hence, we also have $\mathbb{E}(X_t) = t^H \mathbb{E}(X_1) = 0$, for any $t \geq 0$. Being a centered Gaussian process with covariance function defined by equation [2.1], X is a fBm. □

Now, let us introduce complex-valued stochastic processes, Gaussian processes and complex-valued fBm and sfBm.

DEFINITION 2.4.– *Complex-valued stochastic process* $X = \{X_t, t \in T\}$ *is a stochastic process of the form*

$$X_t = X_t^{(1)} + iX_t^{(2)},$$

where $X^{(1)}$ *and* $X^{(2)}$ *are two real-valued stochastic processes,* $i^2 = -1$. *The processes* $X^{(1)}$ *and* $X^{(2)}$ *are called the real and imaginary parts of* X, *correspondingly.*

It is usually said that complex-valued processes have a certain property, if their real and imaginary parts possess this property. For example, we say that a complex-valued process is self-similar and/or has stationary increments, if both parts are self-similar with the same index of self-similarity and/or have stationary increments. However, the situation with Gaussian property is as follows.

DEFINITION 2.5.– *Complex-valued stochastic process* $X = \{X_t, t \in T\}$ *is called a Gaussian process if the vector-valued process* $(X^{(1)}, X^{(2)})$, *where* $X^{(1)}$ *and* $X^{(2)}$ *are real and imaginary parts of* X, *respectively, is a Gaussian process.*

Bearing in mind the above definition, we can now define complex-valued fBm, and the definition of sfBm is formulated similarly.

DEFINITION 2.6.– *Complex-valued Gaussian process* $B^H = \{B_t^H, t \in \mathbb{R}\}$ *is called a fBm if it is a zero-mean Gaussian process with covariance function*

$$G(s,t) = \mathbb{E}B_s^H \overline{B_t^H} = 1/2\left(s^{2H} + t^{2H} - |t-s|^{2H}\right).$$

Obviously, lemma 2.9 is still valid for a complex-valued fBm.

Finally, we are now able to give a spectral representation of a complex-valued fractional Brownian motion and of a complex-valued sub-fractional Brownian motion. Corresponding spectral representations for their real and imaginary parts will immediately follow.

THEOREM 2.2.– *For any* $H \in (0,1)$, *consider a stochastic process* X^H *defined via the spectral representation*

$$X_t^H = \frac{1}{C_2(H)} \int_{\mathbb{R}} \frac{\exp(itx) - 1}{ix} |x|^{\frac{1}{2}-H} dW(x), t \in \mathbb{R}, \quad [2.13]$$

where W is a Wiener measure, whose frequency domain is $\{\mathbb{R}\}$. This means that $W(A)$ is a Gaussian random variable, $\mathbb{E}W(A) = 0$ for any Borel set A, $W(A)$ and $W(B)$ are independent for $A \cap B = \emptyset$, $W(A) = W(-A)$ and

$$\mathbb{E}W(A)^2 = \lambda_1(A),$$

λ_1 denotes the Lebesgue measure on $\mathbb{R}$. Let Γ denote Euler's gamma function, and let

$$C_2(H) = \left(\frac{\pi}{H\Gamma(2H)\sin(\pi H)} \right)^{1/2} = \left(-\frac{2\Gamma(2-2H)\cos \pi H}{H(2H-1)} \right)^{1/2}.$$

Then, X^H is a fractional Brownian motion with Hurst parameter H.

PROOF.– Let f_t be the function of the form

$$f_t(x) = \frac{\exp(itx) - 1}{ix} |x|^{\frac{1}{2}-H}.$$

Taking into account that $\left| \dfrac{\exp(itx) - 1}{ix} \right|$ is bounded as $x \to 0$ and behaves like $|x|^{-1}$ as $x \to \pm\infty$, we get that $\int_{\mathbb{R}} |f_t(x)|^2 dx < \infty$. Therefore, X^H is well defined.

Moreover, $\mathbb{E}|X^H(1)|^2 = C_2^{-2}(H) \int_{\mathbb{R}} |f_1(x)|^2 dx = 1$. Indeed, because

$$|e^{ix} - 1|^2 = (\cos x - 1)^2 + \sin^2 x = 2(1 - \cos x) = 4 \sin^2 \frac{x}{2},$$

we have

$$\int_{\mathbb{R}} |f_1(x)|^2 dx = 8 \int_0^\infty \frac{\sin^2 \frac{x}{2}}{x^{2H+1}} dx.$$

For $H = \frac{1}{2}$, according to relation GR3.821 in [GRA 80],

$$\int_{\mathbb{R}} |f_1(x)|^2 dx = 2\pi = C_2^2 \left(\frac{1}{2} \right).$$

If $H \in (0,1) \setminus \left\{ \frac{1}{2} \right\}$, then it follows from relation GR3.823 in [GRA 80] and from equalities $z\Gamma(z) = \Gamma(z+1)$ and $\Gamma(z)\Gamma(1-z) = \dfrac{\pi}{\sin \pi z}$ that

$$\int_{\mathbb{R}} |f_1(x)|^2 dx = -\frac{2\Gamma(2-2H)\cos \pi H}{H(2H-1)} = C_2^2(H).$$

According to lemma 2.9, to get the proof, it is sufficient to establish that the process X^H is H-self-similar and has stationary increments. To verify self-similarity, consider any $a > 0$ and denote ϕ the characteristic function of some finite-dimensional distribution of the process $\{X^H_{at}, t \geq 0\}$. That is,

$$\phi(\theta) = \mathbb{E} \exp \left(i \sum_{j=1}^{n} \theta_j X^H_{at_j} \right),$$

for some $n \geq 1, t_1, \ldots, t_n \in \mathbb{R}_+$ and $\theta = (\theta_1, \ldots, \theta_n) \in \mathbb{R}^n$.

It follows from lemma A.14 that

$$\phi(\theta) = \exp \left(-\frac{1}{2C_2(H)} \int_{\mathbb{R}} \left| \sum_{j=1}^{n} \theta_j \frac{e^{ixat_j} - 1}{ix} x^{\frac{1}{2}-H} \right|^2 dx \right).$$

Changing the variable $u = xa$, we get

$$\phi(\theta) = \exp \left(-\frac{1}{2C_2(H)} \int_{\mathbb{R}} \left| \sum_{j=1}^{n} a^H \theta_j \frac{e^{iut_j} - 1}{iu} u^{\frac{1}{2}-H} \right|^2 du \right).$$

According to lemma A.14, ϕ is also the characteristic function of the process $a^H X^H$. This means that X^H is H-self-similar.

In order to establish the stationarity of the increments, we can apply the same technique as above and prove that for any $n \geq 1$, any $0 \leq t_1 \leq t_2 \leq \cdots \leq t_n$ and any $h > 0$, the random vectors $(X^H_{t_n} - X^H_{t_{n-1}}, \ldots, X^H_{t_2} - X^H_{t_1})$ and $(X^H_{t_n+h} - X^H_{t_{n-1}+h}, \ldots, X^H_{t_2+h} - X^H_{t_1+h})$ have the same characteristic function. □

REMARK 2.7.– Obviously, real and imaginary parts of the complex-valued process defined via equality [2.13] are the fBms, up to a constant multiplier $1/2$.

REMARK 2.8.– As was established in [DZH 05], we have an inverse statement to theorem 2.2. More precisely, any complex-valued fractional Brownian motion B^H with Hurst parameter H admits a spectral representation [2.13]. Indeed, according to [SAM 94], the covariance function of a complex-valued fBm admits the following representation:

$$\begin{aligned} C(s,t) &= \mathbb{E}(B^H_s \overline{B^H_t}) \\ &= \frac{1}{C_2^2(H)} \int_{\mathbb{R}} \frac{(e^{ixs} - 1)(e^{-ixt} - 1)}{|x|^{1+2H}} d\lambda_1(x) \\ &= \int_{\mathbb{R}} \frac{(e^{ixs} - 1)(e^{-ixt} - 1)}{|x|^2} \mu(dx), \end{aligned}$$

where $\mu(dx) = \frac{1}{C_2^2(H)}|x|^{1-2H}d\lambda_1(x)$ is the so-called spectral measure of fBm. Then, we can apply the Karhunen theorem (see theorem A.13), according to which any fBm B^H can be presented as

$$B_t^H = \int_{\mathbb{R}} \frac{e^{ixs}-1}{ix} \mathcal{Z}(dx),$$

where $\mathcal{Z}$ is an orthogonal random measure with spectral measure μ that is absolutely continuous with respect to the Lebesgue measure λ_1. It immediately follows that

$$\mathcal{Z}(A) = \frac{1}{C_2(H)} \int_A |x|^{1/2-H} dW(x),$$

and we obtain representation [2.13].

COROLLARY 2.3.– *For any $H \in (0,1)$, a real-valued sfBm ξ^H admits the following spectral representation:*

$$\xi_t^H = \frac{\sqrt{2}}{C_2(H)} \int_{\mathbb{R}} \frac{\cos(tx)-1}{x} |x|^{\frac{1}{2}-H} dW(x), \qquad [2.14]$$

where W and $C_2(H)$ are the same as in theorem 2.2.

PROOF.– Equality [2.14] is a direct consequence of equalities [2.8] and [2.13]. □

2.6. Asymptotic growth of a Gaussian self-similar process with application to sub-fractional Brownian motion

Let $X = \{X_t, t \geq 0\}$ be a Gaussian process, self-similar with index $H \in (0,1)$ and with continuous trajectories. Our goal is to establish the rate of asymptotic growth of its trajectories at infinity with probability 1. The key point here is:

THEOREM 2.3.– *Let $X = \{X_t, t \geq 0\}$ be a Gaussian process, self-similar with index $H \in (0,1)$ and with continuous trajectories. Then, for any $t \geq 1$ and any $p > 0$*

$$\sup_{1 \leq s \leq t} |X_s| \leq t^H (\log t + 1)^p \, \xi(p), \qquad [2.15]$$

where $\xi(p)$ is a random variable having some exponential moment, that is, there exists $\varepsilon > 0$ such that

$$\mathbb{E} \exp(\varepsilon \xi(p)) < \infty.$$

PROOF.– Let $p > 0$ be fixed. Denote

$$f(t) = t^H (\log t + 1)^p, \; t \geq 1.$$

Then, f is an increasing function, and moreover,

$$2^{-H}\left(\log 2+1\right)^{-p} f(2t) \leq f(t) \leq f(2t).$$

Therefore, for any $n \geq 1$

$$L_n := \sup_{2^{n-1}\leq s\leq 2^n} \frac{|X_s|}{s^H\left(\log s+1\right)^p} \leq 2^H(\log 2+1)^p \frac{\sup_{2^{n-1}\leq s\leq 2^n}|X_s|}{2^{nH}(n\log 2+1)^p}$$
$$\leq 2^H\left(\frac{\log 2+1}{\log 2}\right)^p \frac{\sup_{2^{n-1}\leq s\leq 2^n}|X_s|}{2^{nH}n^p}.$$

In other words, if we denote $M_n = \frac{\sup_{2^{n-1}\leq s\leq 2^n}|X_s|}{2^{nH}n^p}$, then $L_n \leq C_{p,H}M_n$, where $C_{p,H} = 2^H\left(\frac{\log 2+1}{\log 2}\right)^p$. Now, consider a random variable

$$\xi(p) = \sup_{s\geq 1}\frac{|X_s|}{s^H(\log s+1)^p}.$$

For any $k > 0$, it admits the following upper bound

$$\xi^k(p) \leq \sum_{n=1}^{\infty} L_n^k \leq C_{p,H}^k \sum_{n=1}^{\infty} M_n^k.$$

According to the theorem's conditions, Gaussian process X is continuous on any interval $[0, 2^n]$. It implies that $\sup_{s\in[0,2^n]}|X_s| < \infty$ a.s. Then, it follows from Fernique's theorem, see theorem A.15 and corollary A.7, that for any $k > 0$ $\mathbb{E}\left(\sup_{s\in[0,2^n]}|X_s|^k\right) < \infty$. Furthermore,

$$\mathbb{E}\xi^k(p) \leq C_{p,H}^k \sum_{n=1}^{\infty}\mathbb{E}M_n^k \leq C_{p,H}^k \sum_{n=1}^{\infty}\mathbb{E}\left(\frac{\sup_{s\in[0,2^n]}|X_s|}{2^{nH}}\right)^k n^{-kp}.$$

According to the self-similar property of X, we can conclude that

$$\mathbb{E}\left(\frac{\sup_{s\in[0,2^n]}|X_s|}{2^{nH}}\right)^k = \mathbb{E}(\sup_{s\in[0,1]}|X_s|^k).$$

Therefore,

$$\mathbb{E}\xi^k(p) \leq C_{p,H}^k\mathbb{E}(\sup_{s\in[0,1]}|X_s|^k)\sum_{n=1}^{\infty} n^{-kp}.$$

Choose $k > 1/p$. Then, the series $\sum_{n=1}^{\infty} n^{-kp}$ converges, and that means that $\mathbb{E}\xi^k(p) < \infty$. Consequently, $\xi(p) < \infty$ a.s. This means that

$$|X_s| \leq s^H (\log s + 1)^p \xi(p),$$

whence equation [2.15] follows. Additionally, we note that $\xi(p)$ is also a supremum of absolute values of a centered Gaussian process, and so by virtue of Fernique's theorem (see corollary A.7), the proof follows. □

2.7. Wiener integration with respect to sub-fractional Brownian motion

2.7.1. *Wiener integration with respect to a Gaussian process*

Let $X = \{X_t, t \in [0, T]\}$ be a centered Gaussian process with covariance function $C = \{C(t, s), (t, s) \in [0, T]^2\}$. We assume that the covariance function C is absolutely continuous with respect to the Lebesgue measure λ_2 on $[0, T]^2$. This means that there exists a function $r = \{r(t, s), (t, s) \in [0, T]^2\}$ such that for any $0 \leq s_1 \leq t_1 \leq T$ and $0 \leq s_2 \leq t_2 \leq T$,

$$C(t_1, t_2) - C(t_1, s_2) - C(s_1, t_2) + C(s_1, s_2) = \int_{s_1}^{t_1} \int_{s_2}^{t_2} r(u, v) du dv.$$

Now, let $f : [0, T] \to \mathbb{R}$ be an elementary non-random function, i.e.

$$f(t) = \sum_{k=1}^{n} f_{k-1} \mathbf{1}_{(t_{k-1}, t_k]}, \quad f_k \in \mathbb{R}, \ 0 = t_0 < t_1 < \cdots < t_n = T. \qquad [2.16]$$

DEFINITION 2.7.– *The integral of an elementary function f defined by equality* [2.16], *with respect to a Gaussian process X on $[0, T]$, is defined by*

$$\int_0^T f(t) dX_t = \sum_{k=1}^{n} f_{k-1}(X_{t_k} - X_{t_{k-1}}).$$

Obviously, for any elementary function f, integral $\int_0^T f(t) dX_t$ is a Gaussian random variable with zero mean and variance

$$\begin{aligned}\mathbb{E}\Big(\int_0^T f(t) dX_t\Big)^2 &= \sum_{j=1}^{n} \sum_{k=1}^{n} f_{j-1} f_{k-1} \mathbb{E}\Big(X_{t_j} - X_{t_{j-1}})(X_{t_k} - X_{t_{k-1}})\Big) \\ &= \sum_{j=1}^{n} \sum_{k=1}^{n} f_{j-1} f_{k-1} \Big(C(t_k, t_j) - C(t_{k-1}, t_j) - C(t_k, t_{j-1})\end{aligned}$$

$$+C(t_{k-1},t_{j-1})\Big) = \sum_{j=1}^{n}\sum_{k=1}^{n} f_{j-1}f_{k-1}\int_{t_{k-1}}^{t_k}\int_{t_{j-1}}^{t_j} r(u,v)dudv$$

$$= \int_0^T\int_0^T f(u)f(v)r(u,v)dudv.$$

In this connection, it is natural to introduce the following class of functions.

DEFINITION 2.8.– *Introduce the class of functions*

$$\mathcal{L}([0,T]^2, rd\lambda_2) = \Big\{f : [0,T] \mapsto \mathbb{R} | \int_0^T\int_0^T |f(u)||f(v)||r(u,v)|dudv < \infty\Big\}.$$

The next result guarantees the existence of the integral $\int_0^T f(t)dX_t$, for any function $f \in \mathcal{L}([0,T]^2, rd\lambda_2)$, in the following sense.

THEOREM 2.4 ([MIS 17]).– *Let $f \in \mathcal{L}([0,T]^2, rd\lambda_2)$. Then, there exists a sequence of elementary functions $\{f_n : [0,T] \mapsto \mathbb{R}, n \geq 1\}$ such that the limit of $\int_0^T f_n(t)dX_t$ exists in $L^2(\Omega, \mathcal{F}, \mathbb{P})$, and we can define the integral $\int_0^T f(t)dX_t$ as a limit of the integrals $\int_0^T f_n(t)dX_t$ in $L^2(\Omega, \mathcal{F}, \mathbb{P})$.*

REMARKS 2.8.–

1) Let f be a continuous function on $[0,T]$ and let $\int_0^T\int_0^T |r(u,v)|dudv < \infty$. Then, since f is bounded in $[0,T]$, obviously $f \in \mathcal{L}([0,T]^2, rd\lambda_2)$ and consequently $\int_0^T f(t)dX_t$ exists.

2) Let $f \in \mathcal{L}([0,T]^2, rd\lambda_2)$. Then, $\int_0^T f(t)dX_t$ is a limit in $L^2(\Omega, \mathcal{F}, \mathbb{P})$ of Gaussian random variables with zero mean and variance

$$\int_0^T\int_0^T f_n(u)f_n(v)r(u,v)dudv.$$

Therefore, $\int_0^T f(t)dX_t$ is a zero-mean Gaussian random variable. Moreover, it was proved in theorem 3.6, [MIS 17], that for $f \in \mathcal{L}([0,T]^2, rd\lambda_2)$,

$$\mathbb{E}\Big(\int_0^T f(t)dX_t\Big)^2 = \int_0^T\int_0^T f(u)f(v)r(u,v)dudv.$$

3) We cannot apply theorem 2.4 to a Wiener process W because its covariance function $C(s,t) = s \wedge t$ is not absolutely continuous with respect to the measure λ_2, regardless of the rectangle $[0,T]^2$ considered. Therefore, for the construction of the

integral $\int_0^T f(t)dW_t$, we use the orthogonality of its increments. Let $f \in L^2([0,T])$. Then, there exists a sequence of elementary functions $\{f_n, n \geq 1\}$ such that

$$\lim_{n\to\infty} \|f - f_n\|_{L^2([0,T],\lambda_1)} = 0.$$

Denoting $\int_0^T f_n(t)dW_t = \sum_{k=1}^{n} f_{k-1}(W_{t_k} - W_{t_{k-1}})$, we observe that

$$\mathbb{E}\Big(\int_0^T f_n(t)dW_t\Big) = 0; \text{ and } \mathbb{E}\Big(\int_0^T f_n(t)dW_t\Big)^2 = \sum_{k=1}^{n} f_{k-1}^2(t_k - t_{k-1}),$$

due to the orthogonality of its increments. Indeed, the increments are orthogonal because

$$\mathbb{E}\left((W_{t_k} - W_{t_{k-1}})(W_{t_j} - W_{t_{j-1}})\right) = 0 \text{ for any } j \neq k.$$

Then, for any $n, m \geq 1$

$$\mathbb{E}\left(\int_0^T (f_n(t) - f_m(t))dW_t\right)^2 = \|f_n - f_m\|_{L^2([0,T],\lambda_1)},$$

that is, the isometry property, and having a Cauchy sequence $\{\int_0^T f_n(t)dW_t, n \geq 1\}$, we can define $\int_0^T f(t)dW_t$ as the limit of $\int_0^T f_n(t)dW_t$ in $L^2(\Omega, \mathcal{F}, \mathbb{P})$. Then,

$$\mathbb{E}\Big(\int_0^T f(t)dW_t\Big) = 0 \text{ and } \mathbb{E}\Big(\int_0^T f(t)dW_t\Big)^2 = \int_0^T f^2(t)dt.$$

See also exercise 1.6, where the same integral is introduced but in a slightly different way.

2.7.2. *Wiener integration with respect to fractional and sub-fractional Brownian motion*

Let $\xi^H = \{\xi_t^H, t \geq 0\}$ be a sub-fractional Brownian motion with Hurst parameter $H \in (0,1)$. Recall that its covariance function has the form

$$C(s,t) = s^{2H} + t^{2H} - \frac{1}{2}(s+t)^{2H} - \frac{1}{2}|t-s|^{2H}.$$

Therefore,

$$r(s,t) = \frac{\partial^2 C(s,t)}{\partial s \partial t} = H(2H-1)\Big(|t-s|^{2H-2} - (s+t)^{2H-2}\Big). \qquad [2.17]$$

i) Assume that $H > \frac{1}{2}$.

According to equation [2.17], C is absolutely continuous with respect to the Lebesgue measure λ_2 on any rectangle $[0,T]^2$. Obviously, any bounded measurable function $f : [0,T] \mapsto \mathbb{R}$ belongs to $\mathcal{L}([0,T]^2, rd\lambda_2)$. Therefore, it follows from theorem 2.4 that there exists the integral $\int_0^T f(t) d\xi_t^H$. For fractional Brownian motion B^H with $H > \frac{1}{2}$, we have that

$$C(s,t) = \frac{1}{2}\left(s^{2H} + t^{2H} - |t-s|^{2H}\right),$$

and consequently,

$$r(s,t) = \frac{\partial^2 C(s,t)}{\partial s \partial t} = H(2H-1)|t-s|^{2H-2}.$$

Again, $\int_0^T f(t) dB_t^H$ exists for $f \in \mathcal{L}([0,T]^2, rd\lambda_2)$. In particular, $\int_0^T f(t) dB_t^H$ exists for any bounded measurable function f, and according to [MIS 08],

$$\mathbb{E}\Big(\int_0^T f(t) dB_t^H\Big)^2 \leq C_H \|f\|^2_{L^{1/H}([0,T],\lambda_1)}, \qquad [2.18]$$

where $C_H > 0$ is some constant depending only on H. Moreover, using theorem A.3, inequality [2.18] can be extended to functions $f : \mathbb{R} \to \mathbb{R}$. More precisely, it holds that

$$\mathbb{E}\left(\int_{\mathbb{R}} f(t) dB_t^H\right)^2 \leq C_H \|f\|^2_{L^{1/H}(\mathbb{R},\lambda_1)}.$$

Using a similar approach, we can prove inequality [2.18] for a sub-fractional Brownian motion as well.

THEOREM 2.5.– *Let ξ^H be a sfBm with Hurst parameter $H > \frac{1}{2}$, and a function $f \in \mathcal{L}([0,T]^2, rd\lambda_2)$, with*

$$r(s,t) = \frac{\partial^2 C(s,t)}{\partial s \partial t} = H(2H-1)\Big(|t-s|^{2H-2} - (s+t)^{2H-2}\Big).$$

Then, there exists a constant $C_H > 0$ such that

$$\mathbb{E}\Big(\int_0^T f(t)d\xi_t^H\Big)^2 \leq C_H \|f\|^2_{L^{1/H}([0,T],\lambda_1)}.$$

PROOF.– According to theorem A.3, for any $\alpha \in (0,1)$, $1 < \rho < \frac{1}{\alpha}$ and $q = \frac{p}{1-\alpha p}$, there exists a constant $C_{p,q,\alpha}$ such that

$$\left(\int_{\mathbb{R}}\Big(\int_{\mathbb{R}} |g(u)||x-u|^{\alpha-1}du\Big)^q dx\right)^{1/q} \leq C_{p,q,\alpha}\|g\|_{L^p(\mathbb{R},\lambda_1)}. \qquad [2.19]$$

Now, consider $f : [0,T] \to \mathbb{R}$ and put in equation [2.19]

$$g(u) = \begin{cases} f(-u) \text{ if } u \in [-T,0], \\ \quad 0 \text{ if } u \notin [-T,0]. \end{cases}$$

Then,

$$\int_{\mathbb{R}} |g(u)||x-u|^{\alpha-1}du = \int_{-T}^0 |g(u)||x-u|^{\alpha-1}du = \int_0^T |f(v)||x+v|^{\alpha-1}dv,$$

where in the last equality, we made the change of variables $v = -u$.

Together with equation [2.19], the last inequality means that

$$\left(\int_0^T\left(\int_0^T |f(u)||x+u|^{\alpha-1}du\right)^q dx\right)^{1/q} \leq C_{p,q,\alpha}\|f\|_{L^p([0,T],\lambda_1)}. \qquad [2.20]$$

Now, we put $\alpha = 2H-1 > 0$, $p = \dfrac{1}{H} < \dfrac{1}{2H-1}$ and $q = \dfrac{1}{1-H}$. Applying the Hölder inequality, [2.19] and [2.20], we get that

$$\begin{aligned} &\mathbb{E}\left(\int_0^T f(t)d\xi_t^H\right)^2 \\ &\leq H(2H-1)\Bigg(\int_0^T\int_0^T |f(u)||f(v)|(u+v)^{2H-2}dudv \\ &+ \int_0^T\int_0^T |f(u)||f(v)||u-v|^{2H-2}dudv\Bigg) \end{aligned}$$

$$\leq H(2H-1)\bigg(\int_0^T |f(u)| \left(\int_0^T |f(v)|(u+v)^{2H-2}dv\right) dv$$
$$+ \int_0^T |f(u)| \left(\int_0^T |f(v)||u-v|^{2H-2}dv\right) dv\bigg)$$
$$\leq H(2H-1)\|f\|_{L^{\frac{1}{H}}([0,T],\lambda_1)} \Bigg(\bigg(\int_0^T \bigg(\int_0^T |f(v)|(u$$
$$+v)^{2H-2}dv\bigg)^{\frac{1}{1-H}} du\bigg)^{1-H}$$
$$+ \left(\int_0^T \left(\int_0^T |f(v)||u-v|^{2H-2}dv\right)^{\frac{1}{1-H}} du\right)^{1-H}\Bigg)$$
$$\leq C_H \|f\|^2_{L^{\frac{1}{H}}([0,T],\lambda_1)}. \qquad \square$$

REMARK 2.9.– Of course, all integrals in theorem 2.5 can be replaced with the integrals over $\mathbb{R}$, and the statement of the theorem will be preserved.

ii) Assume now that $0 < H < \frac{1}{2}$. We can introduce the integral $\int_0^T f(t)d\xi_t^H$ for non-random $f : [0,T] \to \mathbb{R}$ having bounded variation, giving the integral the following meaning:

$$\int_0^T f(t)d\xi_t^H = \xi_T^H f(T) - \int_0^T \xi_t^H df(t). \qquad [2.21]$$

Evidently, the right-hand side of equation [2.21] is correctly defined.

2.7.3. *Sub-fractional Ornstein–Uhlenbeck process*

Let us consider the Langevin equation

$$dX_t = aX_t dt + d\xi_t^H, \quad a \in \mathbb{R}, X\big|_{t=0} = X_0 \in \mathbb{R}, \qquad [2.22]$$

where ξ^H is a sub-fractional Brownian motion. Equation [2.22] is a formal differential presentation of the following stochastic process:

$$X_t = X_0 + a\int_0^t X_s ds + \xi_t^H, \; t \geq 0. \qquad [2.23]$$

Similarly to [CHE 03], it is very easy to check that equation [2.23] has the unique solution of the form

$$X_t = X_0 e^{at} + \int_0^t e^{a(t-s)} d\xi_s^H = X_0 e^{at} + \xi_t^H + a \int_0^t \xi_s^H e^{a(t-s)} ds. \qquad [2.24]$$

Stochastic process of the form in equation [2.24] is called a sub-fractional Ornstein–Uhlenbeck (OU) process.

Let us calculate its expectation and variance. Concerning expectation function, it equals $\mathbb{E}(X_t) = X_0 e^{at}$, that is, it has the same form as the expectation of the standard or the fractional Ornstein–Uhlenbeck process, where a standard OU process has the form

$$X_t = X_0 e^{at} + \int_0^t e^{a(t-s)} dW_s,$$

and a fractional OU process has the form

$$X_t = X_0 e^{at} + \int_0^t e^{a(t-s)} dB_s^H = X_0 e^{at} + B_t^H + a \int_0^t B_s^H e^{a(t-s)} ds.$$

Concerning variance of sub-fractional OU process, it equals

$$\begin{aligned}
\mathrm{Var}(X_t) &= \mathbb{E}\left(\xi_t^H + a \int_0^t \xi_s^H e^{a(t-s)} ds\right)^2 \\
&= \mathbb{E}(\xi_t^H)^2 + 2a \int_0^t \mathbb{E}(\xi_t^H \xi_s^H) e^{a(t-s)} ds \\
&\quad + a^2 \int_0^t \int_0^t \mathbb{E}(\xi_v^H \xi_s^H) e^{2at-as-av} ds dv \\
&= (2 - 2^{2H-1}) t^{2H} + 2a e^{at} \int_0^t \left(t^{2H} + s^{2H} - \frac{1}{2}(s+t)^{2H}\right. \\
&\quad \left. - \frac{1}{2}|t-s|^{2H}\right) e^{-as} ds + a^2 e^{2at} \int_0^t \int_0^t \left(v^{2H} + s^{2H} - \frac{1}{2}(s+v)^{2H}\right. \\
&\quad \left. - \frac{1}{2}|v-s|^{2H}\right) e^{-as-av} ds dv
\end{aligned}$$

$$
\begin{aligned}
&= (2 - 2^{2H-1})t^{2H} + 2t^{2H}e^{at}(1 - e^{-at}) + 2ae^{at}\int_0^t e^{-as}s^{2H}ds \\
&\quad - ae^{at}\int_0^t (s+t)^{2H}e^{-as}ds - ae^{at}\int_0^t z^{2H}e^{-a(t-z)}dz \\
&\quad + 2ae^{at}(1 - e^{-at})\int_0^t s^{2H}e^{-as}ds \\
&\quad - \frac{1}{2}a^2e^{2at}\int_0^t\int_0^t \left((s+v)^{2H} + |v-s|^{2H}\right)e^{-as-av}dsdv \\
&= (2e^{at} - 2^{2H-1})t^{2H} - ae^{2at}\int_t^{2t} e^{-az}z^{2H}dz + 2ae^{2at}\int_0^t s^{2H}e^{-as}ds \\
&\quad - a\int_0^t z^{2H}e^{az}dz - \frac{1}{2}a^2e^{2at}\int_0^t\int_0^t \left((s+v)^{2H} + |v-s|^{2H}\right)e^{-as-av}dsdv.
\end{aligned}
$$

Therefore,

$$
\begin{aligned}
\mathrm{Var}(X_t) = {} & (2e^{at} - 2^{2H-1})t^{2H} + 2ae^{2at}\int_0^t s^{2H}e^{-as}ds \\
& - ae^{2at}\int_t^{2t} e^{-az}z^{2H}dz - a\int_0^t s^{2H}e^{as}ds \\
& - \frac{1}{2}a^2e^{2at}\int_0^t\int_v^{t+v} z^{2H}e^{-az}dzdv - a^2e^{2at}\int_0^t e^{-2as}\int_0^s z^{2H}e^{az}dzds.
\end{aligned}
\qquad [2.25]
$$

Let us analyze the asymptotic behavior of the right-hand side of equation [2.25] for different a and H. If $a = 0$, then

$$
\mathrm{Var}(X_t) = (2 - 2^{2H-1})t^{2H} = \mathbb{E}(\xi_t^H)^2,
$$

as expected.

i) Let $a < 0$. Then, we have the following relations:

$$
\lim_{t\to\infty} 2e^{at}t^{2H} = 0 \text{ and } \lim_{t\to\infty} 2ae^{2at}\int_0^t e^{-as}s^{2H}ds = 0, \qquad [2.26]
$$

where the last limit is evaluated via L'Hôpital's rule. Again, according to L'Hôpital's rule,

$$\begin{aligned}
&\lim_{t\to\infty}\Big(-2^{2H-1}t^{2H}-a\frac{\int_t^{2t}z^{2H}e^{-az}dz}{e^{-2at}}-\frac{a^2}{2}\frac{\int_0^t\int_v^{t+v}z^{2H}e^{-az}dzdv}{e^{-2at}}\Big)\\
&=\lim_{t\to\infty}\Big(-2^{2H-1}t^{2H}-a\frac{2(2t)^{2H}e^{-2at}-t^{2H}e^{-at}}{-2ae^{-2at}}\\
&\quad-\frac{a^2}{2}\frac{\int_t^{2t}z^{2H}e^{-az}dz+\int_0^t(t+v)^{2H}e^{-a(t+v)}dv}{-2ae^{-2at}}\Big)\\
&=\lim_{t\to\infty}\Big(-2^{2H-1}t^{2H}-a\frac{2(2t)^{2H}e^{-2at}-t^{2H}e^{-at}}{-2ae^{-2at}}\\
&\quad-\frac{a^2}{2}\frac{2(2t)^{2H}e^{-2at}-t^{2H}e^{-at}}{4a^2e^{-2at}}-\frac{a^2}{2}\frac{\int_t^{2t}z^{2H}e^{-az}dz}{-2ae^{-2at}}\Big)\\
&=\lim_{t\to\infty}\Big(-2^{2H-1}t^{2H}+(2t)^{2H}-\frac{1}{2}t^{2H}e^{at}\\
&\quad-\frac{1}{8}(2.2^{2H}t^{2H}-t^{2H}e^{at})+\frac{1}{8}(-2(2t)^{2H}+t^{2H}e^{at})\Big)\\
&=-\lim_{t\to\infty}\frac{1}{4}t^{2H}e^{at}=0.
\end{aligned}\qquad [2.27]$$

Finally, we apply L'Hôpital's rule again to conclude that

$$\begin{aligned}
&\lim_{t\to\infty}a^2e^{2at}\int_0^t e^{-2as}\int_0^s z^{2H}e^{az}dzds\\
&=\lim_{t\to\infty}a^2\frac{\int_0^t e^{-2as}\int_0^s z^{2H}e^{az}dzds}{e^{-2at}}=-\frac{a}{2}\int_0^{+\infty}z^{2H}e^{az}dz.
\end{aligned}\qquad [2.28]$$

Combining equations [2.25]–[2.28], we get that for $a<0$,

$$\lim_{t\to\infty}\mathrm{Var}(X_t)=-a\int_0^\infty e^{as}s^{2H}ds+\frac{a}{2}\int_0^\infty e^{as}s^{2H}ds=-\frac{a}{2}\int_0^\infty e^{as}s^{2H}ds.$$

Note that for $H=\frac{1}{2}$, the value

$$-\frac{a}{2}\int_0^\infty e^{as}s^{2H}ds=-\frac{a}{2}\int_0^\infty e^{as}sds=-\frac{1}{2a},$$

coincides with the standard formula for OU process.

ii) Let $a > 0$, $H \in \left(0, \frac{1}{2}\right)$. In this case, for any $0 \leq s \leq t$, we have that

$$s^{2H} + t^{2H} \geq (s+t)^{2H}.$$

Therefore,

$$\begin{aligned}\mathbb{E}(\xi_t^H \xi_s^H) &= t^{2H} + s^{2H} - \frac{1}{2}(t+s)^{2H} - \frac{1}{2}(t-s)^{2H} \\ &\geq \frac{1}{2}\left((s+t)^{2H} - (t-s)^{2H}\right) \geq 0.\end{aligned}$$

This means that all terms in the representation

$$\begin{aligned}\mathrm{Var}(X_t) &= \mathbb{E}(\xi_t^H)^2 + 2a\int_0^t \mathbb{E}(\xi_t^H \xi_s^H) e^{a(t-s)} ds \\ &+ a^2 \int_0^t \int_0^t \mathbb{E}(\xi_v^H \xi_s^H) e^{2at-as-av} dsdv\end{aligned}$$

are non-negative, and at least

$$\lim_{t\to\infty} \mathbb{E}(\xi_t^H)^2 = \lim_{t\to\infty} (2 - 2^{2H-1}) t^{2H} = \infty.$$

Therefore, $\lim_{t\to\infty} \mathrm{Var}(X_t) = \infty$.

iii) Now, let $a > 0$ and $H \in \left(\frac{1}{2}, 1\right)$. In this case, we can present X_t in a simpler form

$$X_t = \int_0^t e^{a(t-s)} d\xi_s^H + X_0 e^{at},$$

and deduce from here that

$$\mathrm{Var}(X_t) = H(2H-1)\int_0^t \int_0^t e^{2at-as-av}\left(|v-s|^{2H-2} - (v+s)^{2H-2}\right) dsdv.$$

Note that

$$|v-s|^{2H-2} \geq (v+s)^{2H-2}.$$

Moreover, changing the variables $v = tu$, $s = tz$, we get that

$$\mathrm{Var}(X_t) = H(2H-1)t^{2H}\int_0^1 \int_0^1 e^{at(2-u-z)}\left(|u-z|^{2H-2} - (u+z)^{2H-2}\right) dudz. \quad [2.29]$$

We see that in the right-hand side of equation [2.29]

$$\lim_{t\to\infty} t^{2H} = \infty, e^{at(2-u-z)} > e^{a(2-u-z)}$$

for $t > 1$, and

$$|u-z|^{2H-2} > |u+z|^{2H-2}$$

a.e. with respect to the Lebesgue measure on $[0,1]$. Therefore,

$$\lim_{t\to\infty} \mathrm{Var}(X_t) = \infty.$$

2.8. Compact interval representations of fractional processes

2.8.1. *Fractional integrals and fractional derivatives*

Let interval $[0,T]$ be fixed and consider functions defined on this interval. First, introduce the fractional integrals.

DEFINITION 2.9.– *Let $\alpha > 0$. The right-sided Riemann–Liouville fractional integral operator of order α over $(0,T)$ is defined by*

$$(I^{\alpha}_{T-}f)(s) := \frac{1}{\Gamma(\alpha)} \int_s^T f(t)(t-s)^{\alpha-1}dt, \ s \in (0,T).$$

DEFINITION 2.10.– *Let $\alpha > 0$, $\sigma, \eta \in \mathbb{R}$. The right-sided Erdélyi–Kober-type fractional integral of order α over $(0,T)$ is defined by*

$$(I^{\alpha}_{T-,\sigma,\eta}f)(s) := \frac{\sigma s^{\sigma\alpha}}{\Gamma(\alpha)} \int_s^T (t^{\sigma}-s^{\sigma})^{\alpha-1} t^{\sigma(1-\alpha-\eta)-1} f(t)dt.$$

Second, introduce the fractional derivatives.

DEFINITION 2.11.– *Let $\alpha \in (0,1)$. The right-sided Riemann–Liouville fractional derivative operator of order α over $(0,T)$ is defined by*

$$\begin{aligned}(D^{\alpha}_{T-}f)(s) &:= -\frac{d}{ds}(I^{1-\alpha}_{T-}f)(s)\\ &= -\frac{1}{\Gamma(1-\alpha)}\frac{d}{ds}\int_s^T f(t)(t-s)^{-\alpha}dt.\end{aligned}$$

DEFINITION 2.12.– *Let $0 < \alpha < 1$, $\sigma, \eta \in \mathbb{R}$. The right-sided Erdélyi–Kober-type fractional derivative of order α over $(0,T)$ is defined by*

$$(D^{\alpha}_{T-,\sigma,\eta}f)(s) := s^{\sigma\eta}\left(-\frac{d}{\sigma s^{\sigma-1}ds}\right)\left(s^{\sigma(1-\eta)}(I^{-\alpha+1}_{T-,\sigma,\eta-1}f)\right)(s), \ s \in (0,T).$$

Fractional Riemann–Liouville integrals and derivatives participate in the transformations of fractional Brownian motion, while the Erdélyi–Kober-type fractional integrals and derivatives participate in the respective transformations of sub-fractional Brownian motion. In order to consider these transformations, introduce the following kernels. Namely, denote

$$F^H(t,s) = C_{H,1} s^{\frac{1}{2}-H} \left(I_{t-}^{\frac{1}{2}-H} u^{H-\frac{1}{2}} f(u) \right)(s),\ s \in (0,t),$$

where

$$C_{H,1} = \left(\frac{\Gamma(2-2H)}{2H\Gamma(H+\frac{1}{2})\Gamma(\frac{3}{2}-H)} \right)^{\frac{1}{2}},$$

and

$$(SF)^H(t,s) = C_{H,2} s^{H-\frac{1}{2}} \Big(t^{H-\frac{3}{2}} (t^2-s^2)^{\frac{1}{2}-H} - \left(H-\frac{3}{2}\right) \int_s^t (u^2-s^2)^{\frac{1}{2}-H} u^{H-\frac{3}{2}} du \Big),\ s \in (0,t),$$

where

$$C_{H,2} = \frac{1}{\Gamma(\frac{3}{2}-H)}.$$

For the Riemann–Liouville integrals defined on $\mathbb{R}$ and for the Hardy–Littlewood theorem, see section A.4 in the Appendix.

2.8.2. *Compact interval representation of fBm and sfBm: fundamental martingale*

Let us start with fractional Brownian motion $B^H = \{B_t^H, t \geq 0\}$. It was established in [NOR 99] and [JOS 06] that having kernel $F^H(t,s)$ we can receive a Wiener process as follows.

PROPOSITION 2.4.– *The stochastic process defined as*

$$W_t = \int_0^t F^H(t,s)dB_s^H,\ t \geq 0 \qquad [2.30]$$

is well defined and is a Wiener process generating the same filtration as B^H.

Similarly, according to [DZH 04], we can transform sfBm ξ^H into a Wiener process as follows.

PROPOSITION 2.5.– *The process*

$$W_t = \int_0^t (SF)^H(t,s)d\xi_s^H,\ t \geq 0, \tag{2.31}$$

is well defined and is a Wiener process generating the same filtration as ξ^H.

Formulas of inverse transformation are also well known and have the following form:

PROPOSITION 2.6.–

i) Let the Wiener process be defined by formula [2.30]. Then,

$$B_t^H = \int_0^t G^H(t,s)dW_s,\ t \geq 0,$$

where

$$G^H(t,s) = C_{H,3}\big(t^{H-\frac{1}{2}}s^{\frac{1}{2}-H}(t-s)^{H-\frac{1}{2}} - \left(H-\frac{1}{2}\right)s^{\frac{1}{2}-H}\int_s^t u^{H-\frac{3}{2}}(u-s)^{H-\frac{1}{2}}du\big),$$

and $C_{H,3} = C_{H,1}^{-1}$.

ii) Let the Wiener process be defined by formula [2.31]. Then,

$$\xi_t^H = C_{H,4}\int_0^t (SG)^H(t,s)dW_s,\ t \geq 0,$$

where

$$C_{H,4} = \left(\frac{2H\Gamma(2H)\sin(\pi H)}{\pi}\right)^{\frac{1}{2}}$$

and

$$(SG)^H(t,s) = \frac{\sqrt{\pi}}{2^{H-\frac{1}{2}}} I_{t-,2,\frac{3}{4}-\frac{H}{2}}^{H-\frac{1}{2}}(u^{H-1/2})(s).$$

Sometimes, it is more convenient to use not Wiener process itself, but the Gaussian martingale

$$M_t^H = (2-2H)^{-\frac{1}{2}}\int_0^t s^{\frac{1}{2}-H}dW_s,$$

with bracket $\langle M^H \rangle_t = t^{2-2H}$. This martingale is called a Molchan martingale and it admits a very simple representation via fractional Brownian motion.

PROPOSITION 2.7 ([NOR 99]).– *The following relation holds: for any $H \in (0,1)$, a stochastic process defined as*

$$M_t = C_H \int_0^t (t-s)^{1/2-H} s^{1/2-H} dB_s^H, \; t \geq 0,$$

is a Molchan martingale. Here

$$C_H = \left(\frac{\Gamma(3-2H)}{2H\Gamma^3(3/2-H)\Gamma(H+1/2)} \right)^{1/2}.$$

2.9. Girsanov theorem for sub-fractional Brownian motion

At the outset, recall the well-known standard Girsanov theorem for a Wiener process with a drift. This result allows us to annihilate the drift in the following sense.

PROPOSITION 2.8.– *Let $W = \{W_t, \mathcal{F}_t^W, t \geq 0\}$ be a Wiener process on the probability space $(\Omega, \mathcal{F}, \mathbb{P})$ with respect to its own filtration $\{\mathcal{F}_t^W, t \geq 0\} \subset \mathcal{F}$. Also, let the random function $f = \{f(t,\omega), t \geq 0, \omega \in \Omega\}$ be a real-valued $\mathcal{F}_t^W$-adapted measurable function such that $\int_0^T f^2(s,\omega)ds < \infty$ a.s. If*

$$\mathbb{E} \exp \left(\int_0^T f(t,\omega) dW_t - \frac{1}{2} \int_0^T f^2(t,\omega) dt \right) = 1,$$

then $W_t - \int_0^T f(s,\omega)ds$ is a Wiener process with respect to the probability measure $\mathbb{Q} \sim \mathbb{P}$ such that

$$\left. \frac{d\mathbb{Q}}{d\mathbb{P}} \right|_{\mathcal{F}_T^W} = \exp \left(\int_0^T f(t,\omega) dW_t - \frac{1}{2} \int_0^T f^2(t,\omega) dt \right).$$

Now, we can proceed to fractional processes in order to annihilate the drift for them. The corresponding results are straightforward. Indeed, let us try to find the measure $\mathbb{Q}^H \sim \mathbb{P}$ such that $B_t^H - \int_0^t f(s,\omega)ds$ is a fractional Brownian motion on $[0,T]$ with respect to $\mathbb{Q}^H$. Then, applying transformation formula [2.30], we get that

$$W_t = \int_0^t F^H(t,s) dB_s^H - \int_0^t F^H(t,s) f(s,\omega) ds, \; t \in [0,T]$$

is a Wiener process with respect to the measure $\mathbb{Q}^H$ such that

$$\frac{d\mathbb{Q}^H}{d\mathbb{P}}\bigg|_{\mathcal{F}_T} = \exp\left(\int_0^T \varphi(t,\omega)dW_t - \frac{1}{2}\int_0^T \varphi^2(t,\omega)dt\right),$$

where $\displaystyle\int_0^t F^H(t,s)f(s,\omega)ds = \int_0^t \varphi(s,\omega)ds.$

Therefore, the following result can be established. For the proof, see theorem 2.8.1 in [MIS 08].

THEOREM 2.6.– *Let the following conditions hold: there exists a function* $\varphi : [0,T] \times \Omega \to \mathbb{R}$ *such that*

1) $\displaystyle\int_0^t F^H(t,s)f(s,\omega)ds = \int_0^t \varphi(s,\omega)ds$ *and*

$$\int_0^t \varphi^2(s,\omega)ds < \infty \ a.s., \quad t \in [0,T].$$

2) $\displaystyle\mathbb{E}\exp\left(\int_0^T \varphi(t,\omega)dW_t - \frac{1}{2}\int_0^T \varphi^2(t,\omega)dt\right) = 1.$

Then, $B_t^{\mathbb{Q}^H} := B_t^H - \int_0^t f(s,\omega)ds$ *is a fractional Brownian motion with respect to the measure* $\mathbb{Q}^H$ *such that*

$$\frac{d\mathbb{Q}^H}{d\mathbb{P}}\bigg|_{\mathcal{F}_T} = \exp\left(\int_0^T \varphi(t,\omega)dW_t - \frac{1}{2}\int_0^T \varphi^2(t,\omega)dt\right).$$

Obviously, a similar result holds for sub-fractional Brownian motion, only replacing F^H *for* $(SF)^H$.

2.10. Comparison of fractional and sub-fractional processes via Slepian's lemma

First, let us recall Slepian's lemma, proved in 1962 in [SLE 62].

LEMMA 2.10.– *Let* $X = (X_1,\ldots,X_n)$ *and* $Y = (Y_1,\ldots,Y_n)$ *be two Gaussian vectors satisfying*

i) $\mathbb{E}(X_i) = \mathbb{E}(Y_j) = 0$, $1 \le i,j \le n$,

ii) $\mathbb{E}(X_i^2) = \mathbb{E}(Y_i^2)$, $1 \le i \le n$ *and iii)* $\mathbb{E}(X_iX_j) \leq \mathbb{E}(Y_iY_j)$ *for* $1 \leq i, j \leq n$.

Then, for any real number $u_1, \ldots, u_n$, *the following inequality holds*

$$\mathbb{P}(X_1 \leq u_1, \ldots, X_n \leq u_n) \leq \mathbb{P}(Y_1 \leq u_1, \ldots, Y_n \leq u_n).$$

Now, let us compare fractional Brownian motion B^H and sub-fractional Brownian motion ξ^H using Slepian's lemma. In this connection, we should reduce $\mathbb{E}(\xi_s^H)^2$ to s^{2H} by rationing. Note that

$$\mathbb{E}(\xi_s^H)^2 = (2 - 2^{2H-1})s^{2H}.$$

Therefore, for $H > \frac{1}{2}$, it holds that $\mathbb{E}(\xi_s^H)^2 < \mathbb{E}(B_s^H)^2$, while for $H < \frac{1}{2}$, it holds that $\mathbb{E}(\xi_s^H)^2 > \mathbb{E}(B_s^H)^2$, for any $s > 0$.

According to the above, introduce the normalizing constant $c_H = (2 - 2^{2H-1})^{-1/2}$, and let $\eta_s^H = c_H \xi_s^H$ be a normalized sub-fractional Brownian motion with $\mathbb{E}(\eta_s^H)^2 = s^{2H}$.

Consider, for any $0 \leq s \leq t$, the covariance function of the normalized sfBm

$$C_{\eta^H}(s,t) = \mathbb{E}(\eta_s^H \eta_t^H) = c_H^2 \left(s^{2H} + t^{2H} - \frac{1}{2}(s+t)^{2H} - \frac{1}{2}(t-s)^{2H} \right)$$

and compare it to the covariance function of fBm:

$$C_{B^H}(s,t) = \mathbb{E}(B_s^H B_t^H) = \frac{1}{2}\left(s^{2H} + t^{2H} - (t-s)^{2H} \right).$$

LEMMA 2.11.–

i) For any $H \in \left(0, \frac{1}{2}\right)$ *and* $0 \leq s \leq t$, *we have that* $C_{\eta^H}(s,t) \geq C_{B^H}(s,t)$.

ii) For any $H \in \left(\frac{1}{2}, 1\right)$ *and* $0 \leq s \leq t$, *we have that* $C_{\eta^H}(s,t) \leq C_{B^H}(s,t)$.

PROOF.–

i) Let $H \in \left(0, \frac{1}{2}\right)$.

Our aim is to establish the following relation: for $0 \leq s \leq t$,

$$c_H \left(s^{2H} + t^{2H} - \frac{1}{2}(s+t)^{2H} - \frac{1}{2}(t-s)^{2H} \right) \geq \frac{1}{2}\left(s^{2H} + t^{2H} - (t-s)^{2H} \right), \quad [2.32]$$

or, as is equivalent,

$$\left(s^{2H}+t^{2H}-\frac{1}{2}(s+t)^{2H}-\frac{1}{2}(t-s)^{2H}\right)\geq(1-2^{2H-2})\left(s^{2H}+t^{2H}-(t-s)^{2H}\right).[2.33]$$

Inequality [2.33] is equivalent to the following:

$$2^{2H-2}s^{2H}+2^{2H-2}t^{2H}-\frac{1}{2}(s+t)^{2H}\geq\left(-\frac{1}{2}+2^{2H-2}\right)(t-s)^{2H}. \quad [2.34]$$

Obviously, inequality [2.34] becomes an identity for $s=0$. For $s\neq 0$, we can divide both parts of inequality [2.34] by s^{2H} and get an equivalent inequality

$$\left(\frac{1}{2}-2^{2H-2}\right)(z-1)^{2H}\geq\frac{1}{2}(z+1)^{2H}-2^{2H-2}z^{2H}-2^{2H-2}, \quad [2.35]$$

where $z=\frac{t}{s}\in[1,+\infty)$. Inequality [2.35] holds for $z=1$ because both its sides equal zero. The derivative of the left-hand side equals

$$\left(\frac{1}{2}-2^{2H-2}\right)2H(z-1)^{2H-1}=H\left(1-2^{2H-1}\right)(z-1)^{2H-1},$$

while the derivative of the right-hand side equals

$$H(z+1)^{2H-1}-2^{2H-1}Hz^{2H-1}.$$

In order to establish inequality [2.35], it is sufficient to prove that

$$(1-2^{2H-1})(z-1)^{2H-1}\geq(z+1)^{2H-1}-2^{2H-1}z^{2H-1},\quad z\in[1,+\infty).[2.36]$$

Let us divide both parts of equation [2.36] by z^{2H-1}, denote $v=\frac{1}{z}$ and get an equivalent inequality

$$(1-2^{2H-1})(1-v)^{2H-1}\geq(1+v)^{2H-1}-2^{2H-1},\quad v\in(0,1]. \quad [2.37]$$

Because $2H-1<0$, we have that the left-hand side of equation [2.37] increases in v from $1-2^{2H-1}$ to $+\infty$, while the right-hand side decreases in v from $1-2^{2H-1}$ to 0. This means that equation [2.37] is true. Therefore, all of inequalities [2.32]–[2.36] are true as well, and we get the proof for $H\in\left(0,\frac{1}{2}\right)$.

ii) Let $H \in \left(\frac{1}{2}, 1\right)$. Our goal is to prove that for $0 \leq s \leq t$

$$c_H\left(s^{2H} + t^{2H} - \frac{1}{2}(s+t)^{2H} - \frac{1}{2}(t-s)^{2H}\right) \leq \frac{1}{2}\left(s^{2H} + t^{2H} - (t-s)^{2H}\right).$$

Applying the same transformations as were above applied to [2.32]–[2.34], we conclude that it is sufficient to establish the relation

$$\left(\frac{1}{2} - 2^{2H-2}\right)(z-1)^{2H} \leq \frac{1}{2}(z+1)^{2H} - 2^{2H-2}z^{2H} - 2^{2H-2}, \qquad [2.38]$$

where $z \in [1, +\infty)$. Again, equation [2.38] holds for $z = 1$. Let us compare the derivatives of the left- and right-hand sides of equation [2.38]. To achieve our goal, it is sufficient to prove that

$$(1 - 2^{2H-1})(z-1)^{2H-1} \leq (z+1)^{2H-1} - 2^{2H-1}z^{2H-1}. \qquad [2.39]$$

Transforming equation [2.39] as was done for equation [2.36], we get the relation

$$(1 - 2^{2H-1})(1-v)^{2H-1} \leq (1+v)^{2H-1} - 2^{2H-1}, \quad v \in (0, 1].$$

or, equivalently,

$$(2^{2H-1} - 1)(1-v)^{2H-1} \geq 2^{2H-1} - (1+v)^{2H-1}, \quad v \in (0, 1]. \qquad [2.40]$$

Relation [2.40] holds for $v = 0$ and $v = 1$. Indeed, for $v = 0$, both of its parts equal $2^{2H-1} - 1$, and for $v = 1$, both of its parts equal zero. Therefore, the function

$$g(v) = (2^{2H-1} - 1)(1-v)^{2H-1} - 2^{2H-1} + (1+v)^{2H-1}$$

equals zero at 0 and at 1. Its second derivative

$$g''(v) = (2H-1)(2H-2)\left[(1+v)^{2H-3} + (2^{2H-1} - 1)(1-v)^{2H-3}\right] < 0$$

on the interval $(0, 1)$. This means that g is concave and therefore $g(v) > 0$ on $(0, 1)$. The lemma is proved. □

THEOREM 2.7.– *Consider two Gaussian processes:*

$$\eta^H = \{\eta_t^H = \left(2 - 2^{2H-1}\right)\xi_t^H, t \geq 0\} \text{ and } B^H = \{B_t^H, t \geq 0\},$$

which are normalized sub-fractional and fractional Brownian motions, respectively.

Then,

i) For $H \in (0, 1/2)$*, any* $0 \leq t_1 \leq t_2 \leq \cdots \leq t_n$ *and any vector* $(x_1, x_2, \ldots, x_n) \in \mathbb{R}^n$*, we have that*

$$\mathbb{P}(B^H_{t_1} \leq x_1, B^H_{t_2} \leq x_2, \ldots, B^H_{t_n} \leq x_n) \\ \leq \mathbb{P}(\eta^H_{t_1} \leq x_1, \eta^H_{t_2} \leq x_2, \ldots, \eta^H_{t_n} \leq x_n).$$

ii) For $H \in (1/2, 1)$*, any* $0 \leq t_1 \leq t_2 \leq \cdots \leq t_n$ *and any vector* $(x_1, x_2, \ldots, x_n) \in \mathbb{R}^n$*, we have that*

$$\mathbb{P}(\eta^H_{t_1} \leq x_1, \eta^H_{t_2} \leq x_2, \ldots, \eta^H_{t_n} \leq x_n) \\ \leq \mathbb{P}(B^H_{t_1} \leq x_1, B^H_{t_2} \leq x_2, \ldots, B^H_{t_n} \leq x_n).$$

Proof of the theorem immediately follows from lemmas 2.10 and 2.11.

COROLLARY 2.4.–

i) For $H \in (0, 1/2)$*,* $T > 0$ *and* $x > 0$

$$\mathbb{P}(\sup_{t\in[0,T]} B^H_t \leq x) \leq \mathbb{P}(\sup_{t\in[0,T]} \eta^H_t \leq x).$$

ii) For $H \in (1/2, 1)$*,* $T > 0$ *and* $x > 0$

$$\mathbb{P}(\sup_{t\in[0,T]} \eta^H_t \leq x) \leq \mathbb{P}(\sup_{t\in[0,T]} B^H_t \leq x).$$

2.11. Exercises

EXERCISE 2.1.– *Consider the so-called Liouville fractional Brownian motion, i.e. the process* $\mathcal{L}^H = \{\mathcal{L}^H_t; t \geq 0\}$ *that admits a representation*

$$\mathcal{L}^H_t = \int_0^t (t-s)^{H-\frac{1}{2}} dW_s, \; t \geq 0, \; H \in (0,1),$$

and that $W = \{W_t, t \geq 0\}$ *is a Wiener process. Prove that* $\mathcal{L}^H$ *is a centered Gaussian process and find its covariance function.*

EXERCISE 2.2.– *Consider a Langevin equation, i.e. a stochastic differential equation*

$$X_t = X_0 + a \int_0^t X_s ds + \sigma B^H_t, \qquad [2.41]$$

where $X_0, a \in \mathbb{R}$*,* $\sigma > 0$*,* B^H *is a fractional Brownian motion,* $H \in (0,1)$*.*

1) Prove that its (unique) solution has a form

$$X_t = X_0 \exp(at) + a \int_0^t \exp(a(t-s)) B_s^H ds + B_t^H.$$

Consider the Langevin equation with sfBm ξ^H instead of fBm B^H and prove the existence–uniqueness result.

2) Prove that X satisfying equation [2.41] is a Gaussian process and find its covariance function. Likewise for sfBm ξ^H instead of fBm B^H.

EXERCISE 2.3.– *Let $B^H = \{B_t^H, t \in \mathbb{R}\}$ be a fractional Brownian motion.*

1) Prove that $\mathbb{E}(B_{-t}^H B_{-s}^H) = \mathbb{E}(B_t^H B_s^H)$ for any $s, t \geq 0$.

2) Prove that the increments of B^H are positively correlated for $H \in \left(\frac{1}{2}, 1\right)$ and negatively correlated for $H \in \left(0, \frac{1}{2}\right)$.

3) Based on the above facts, prove that for sfBm ξ^H, the increments are also positively correlated for $H \in \left(\frac{1}{2}, 1\right)$ and negatively correlated for $H \in \left(0, \frac{1}{2}\right)$.

EXERCISE 2.4.– *Let $B_i^H = \{B_i^H(t), t \in \mathbb{R}\}, 1 \leq i \leq n$, be the independent fBms with the same Hurst index. Prove that any linear combination $\sum_{i=1}^n a_i B_i^H$ is a fractional Brownian motion with the same Hurst index, up to a constant multiplier. Prove the same statement for sfBms.*

EXERCISE 2.5.– *Let $B_i^H = \{B_i^H(t), t \in \mathbb{R}\}, 1 \leq i \leq n$, be the independent fBms with the same Hurst index. Taking some of them, create respective sfBms:*

$$\xi_i^H(t) = \frac{1}{\sqrt{2}} \left(B_i^H(t) + B_i^H(-t)\right), 1 \leq i \leq i_0 < n.$$

Prove that any linear combination $\sum_{i=1}^{i_0} a_i B_i^H + \sum_{i=i_0+1}^n a_i \xi_i^H$ can be presented as

$$aB_t^H + bB_{-t}^H,$$

where B^H is a fractional Brownian motion, $a, b \in \mathbb{R}$.

EXERCISE 2.6.– *Let fBm B^H and sfBm ξ^H be independent stochastic processes, and*

$$\xi_t^H = \frac{1}{\sqrt{2}} (\widehat{B}_t^H + \widehat{B}_{-t}^H), t \geq 0,$$

where $\widehat{B}^H$ *is an fBm. Prove that it does not imply that* B^H *and* $\widehat{B}^H$ *are independent. Hint: analyze the proof of lemma 2.4.*

EXERCISE 2.7.– *Consider a fractional Brownian motion* $B^H = (B_t^H)_{t\in\mathbb{R}}$ *of Hurst parameter* H*, and the even and odd parts of* B^H*, the processes defined respectively by:*

$$B_t^{H,e} = \frac{B_t^H + B_{-t}^H}{2}, \quad B_t^{H,o} = \frac{B_t^H - B_{-t}^H}{2}; \; t \in \mathbb{R}.$$

Note that $B^{H,e}$ *is a sub-fractional Brownian motion, up to a constant multiplier.*

1) Prove that both $B_t^{H,e}$ *and* $B_t^{H,o}$ *are Gaussian processes.*

2) Prove that the sample paths of $B^{H,e}$ *and* $B^{H,o}$ *are, respectively, even and odd functions.*

3) Prove that the Gaussian processes $B^{H,o}$ *and* $B^{H,e}$ *are uncorrelated and consequently independent, respectively.*

3

Mixed Fractional and Mixed Sub-fractional Brownian Motions

In this chapter, the extensions of a fBm and a sfBm are introduced, and will be called a mixed fractional Brownian motion (mfBm) and a mixed sub-fractional Brownian motion (msfBm), respectively.

DEFINITION 3.1.– *Let $(\Omega, \mathcal{F}, \mathbb{P})$ be a probability space, N be a positive integer number, vector of indices $H = (H_1, H_2, \ldots, H_N) \in (0,1)^N$ and vector of coefficients $a = (a_1, a_2, \ldots, a_N) \in \mathbb{R}^N$, with $a_i \neq 0$. A mixed fractional Brownian motion (mfBm) and a mixed sub-fractional Brownian motion (msfBm) that depend on parameters N, a and H are the processes*

$$M = \{M_t^H(N,a),\ t \geq 0\} = \{M_t^H(a),\ t \geq 0\} = \{M_t^H,\ t \geq 0\}$$

and

$$S = \{S_t^H(N,a),\ t \geq 0\} = \{S_t^H(a),\ t \geq 0\} = \{S_t^H,\ t \geq 0\},$$

respectively, defined on $(\Omega, \mathcal{F}, \mathbb{P})$ via the formulas

$$M_t^H(N,a) = \sum_{i=1}^{N} a_i B_t^{H_i} \text{ and } S_t^H(N,a) = \sum_{i=1}^{N} a_i \xi_t^{H_i}, \qquad [3.1]$$

respectively. Here, $\{B^{H_i},\ 1 \leq i \leq N\}$ and $\{\xi^{H_i},\ 1 \leq i \leq N\}$ are the families consisting of independent fractional Brownian motions and sub-fractional Brownian motions, respectively, of Hurst parameters H_i, defined on $(\Omega, \mathcal{F}, \mathbb{P})$.

REMARK 3.1.– Considering the mixed processes, we assume that $N \geq 2$. Taking into account exercise 2.4, we can assume that all H_i are different. Moreover, without loss of generality, assume that $0 < H_1 < H_2 < \ldots < H_N$. Emphasize again that all coefficients a_i are non-zero.

REMARK 3.2.– If we assume exceptionally that for the moment $N = 1$ and $a_1 = 1$, then $M^H = B^H$ ($S^H = \xi^H$) is a fractional Brownian motion (sub-fractional Brownian motion). If $N = 1$, $H_1 = 1/2$ and $a_1 = 1$, then both M^H and S^H are the standard Brownian motions. Therefore, the mfBm and msfBm are obviously more general processes than Bm, fBm and sfBm.

3.1. The main properties of mixed fractional and mixed sub-fractional Brownian motions

LEMMA 3.1.– *The mfBm and the msfBm satisfy the following properties:*

i) $M^H(a)$ *and* $S^H(a)$ *are centered Gaussian processes.*

ii) For any $s, t \in \mathbb{R}_+$,

$$\mathrm{Cov}\left(M_t^H(a), M_s^H(a)\right) = \frac{1}{2}\sum_{i=1}^{N} a_i^2 \left(t^{2H_i} + s^{2H_i} - |t-s|^{2H_i}\right), \qquad [3.2]$$

and

$$\mathrm{Cov}\left(S_t^H(a), S_s^H(a)\right) = \sum_{i=1}^{N} a_i^2 \left(t^{2H_i} + s^{2H_i}\right.$$
$$\left. -1/2\left((s+t)^{2H_i} + |t-s|^{2H_i}\right)\right). \qquad [3.3]$$

COROLLARY 3.1.– *For any* $t \geq 0$

$$\mathbb{E}\left(M_t^H(a)\right)^2 = \sum_{i=1}^{N} a_i^2 t^{2H_i}, \qquad [3.4]$$

and

$$\mathbb{E}\left(S_t^H(a)\right)^2 = \sum_{i=1}^{N} a_i^2 \left((2 - 2^{2H_i - 1})t^{2H_i}\right). \qquad [3.5]$$

Let us now study the mixed-self-similarity property (see [ZIL 06]) of the mfBm and the msfBm.

LEMMA 3.2.– *Both processes* M^H *and* S^H *satisfy the mixed self-similarity property, which is formulated as follows: for any* $h > 0$, *the processes*

$$\left\{S_{ht}^H(a), t \geq 0\right\} \text{ and } \left\{S_t^H\left(a_1 h^{H_1}, a_2 h^{H_2}, \ldots, a_N h^{H_N}\right), t \geq 0\right\},$$

and, respectively,

$$\{M_{ht}^H(a), t \geq 0\} \text{ and } \left\{M_t^H\left(a_1 h^{H_1}, a_2 h^{H_2}, \ldots, a_N h^{H_N}\right), t \geq 0\right\}$$

have the same probability law.

PROOF.– This is due to the fact that for fixed $h > 0$, stochastic processes

$$\{M_{ht}^H(a), t \geq 0\} \quad \text{and} \quad \{M_t^H\left(a_1 h^{H_1}, a_2 h^{H_2}, \ldots, a_N h^{H_N}\right), t \geq 0\}$$

are centered Gaussian processes with the same covariance functions. A similar transformation takes place for S with respective indices. □

The following lemma deals with the non-Markov property of the mfBm and the msfBm.

THEOREM 3.1.– *Let there exist $H_j \neq 1/2$. Then, both $\{M_t^H(a),\ t \geq 0\}$ and $\{S_t^H(a),\ t \geq 0\}$ are not Markov processes.*

PROOF.– Consider only msfBm S^H. We leave the case of mfBm to the reader as an exercise (see exercise 3.1). First, S^H is a centered Gaussian process. Secondly, there exists $j \in \{1, \ldots, N\}$ such that $H_j \neq 1/2$. If the process S^H was Markov, then, according to [REV 91], for all $s < t < u$, the following equality would be correct:

$$\text{Cov}\left(S_s^H, S_u^H\right) \text{Cov}\left(S_t^H, S_t^H\right) = \text{Cov}\left(S_s^H, S_t^H\right) \text{Cov}\left(S_t^H, S_u^H\right). \qquad [3.6]$$

i) Let there exist $H_j > 1/2$. In this case, $H_N > 1/2$, and H_N is the biggest index according to our assumption. Substituting

$$1 < s = \sqrt{t} < t < u = t^2$$

into equalities [3.3] and [3.6], we get that the following equality must hold:

$$\begin{aligned}
&\sum_{i=1}^N a_i^2 (2 - 2^{2H_i - 1}) t^{2H_i} \\
&\quad \times \sum_{i=1}^N a_i^2 \left(t^{4H_i} + t^{H_i} - 1/2 t^{4H_i} \left[(1 + t^{-3/2})^{2H_i} + (1 - t^{-3/2})^{2H_i} \right] \right) \\
&= \sum_{i=1}^N a_i^2 \left(t^{H_i} + t^{2H_i} - 1/2 t^{2H_i} \left[(1 + t^{-1/2})^{2H_i} + (1 - t^{-1/2})^{2H_i} \right] \right) \\
&\quad \times \sum_{i=1}^N a_i^2 \left(t^{2H_i} + t^{4H_i} - 1/2 t^{4H_i} \left[(1 + t^{-1})^{2H_i} + (1 - t^{-1})^{2H_i} \right] \right). \qquad [3.7]
\end{aligned}$$

Note that asymptotically as $h \to 0$ we have

$$(1+h)^{2H_i} + (1-h)^{2H_i} = 2 + 2H_i(2H_i - 1)h^2 + o(h^2). \tag{[3.8]}$$

Therefore, on the one hand, it follows from equalities [3.7] and [3.8] that the limit relation must be satisfied

$$\begin{aligned}
&\sum_{i=1}^{N} a_i^2(2 - 2^{2H_i-1})t^{2H_i} \sum_{i=1}^{N} a_i^2 \left(t^{H_i} - H_i(2H_i-1)t^{4H_i-3} + o\left(t^{4H_i-3}\right)\right) \\
&\quad - \left(\sum_{i=1}^{N} a_i^2 \left(t^{H_i} - H_i(2H_i-1)t^{2H_i-1} + o\left(t^{2H_i-1}\right)\right)\right) \\
&\quad \times \left(\sum_{i=1}^{N} a_i^2 \left(t^{2H_i} - H_i(2H_i-1)t^{4H_i-2} + o\left(t^{4H_i-2}\right)\right)\right) \\
&\quad = a_N^4(1 - 2^{2H_N-1})t^{3H_N} + o(t^{3H_N})),
\end{aligned}$$

as $t \to \infty$. On the other hand, this value should be zero. Such equality can only hold if $H_N = 1/2$. Therefore, S^H is not a Markov process.

ii) Let there exist $H_j < 1/2$. In this case, $H_1 < 1/2$. Let us put in equality [3.7]

$$0 < s = t^2 < t < u = \sqrt{t} < 1.$$

Then, we get that the following equality must hold:

$$\begin{aligned}
&\sum_{i=1}^{N} a_i^2(2 - 2^{2H_i-1})t^{2H_i} \\
&\quad \times \sum_{i=1}^{N} a_i^2 \left(t^{4H_i} + t^{H_i} - 1/2t^{H_i}\left((1+t^{3/2})^{2H_i} + (1-t^{3/2})^{2H_i}\right)\right) \\
&\quad = \sum_{i=1}^{N} a_i^2 \left(t^{2H_i} + t^{4H_i} - 1/2t^{2H_i}\left((1+t)^{2H_i} + (1-t)^{2H_i}\right)\right) \\
&\quad \times \sum_{i=1}^{N} a_i^2 \left(t^{H_i} + t^{2H_i} - 1/2t^{H_i}\left((1+t^{1/2})^{2H_i} + (1-t^{1/2})^{2H_i}\right)\right).
\end{aligned} \tag{[3.9]}$$

The smallest power in equation [3.9] is $6H_1$. Dividing the left- and right-hand sides of equation [3.9] for t^{6H_i}, we get that it should be

$$\lim_{t \to 0} a_1^4(1 - 2^{2H_1-1}) = 0,$$

which is true if and only if $H_1 = 1/2$. Thus, in this case too, S^H is not a Markov process. □

3.2. The behavior of the increments of mfBm and msfBm

Let us recall the notion of quasi-helix, introduced in Chapter 1, definition 1.11. We say that a stochastic process $X = \{X_t, t \in \mathbb{T} \subseteq \mathbb{R}\}$ belongs to the quasi-helix

$$\mathbb{QH}(H, C_1, C_2, \mathbb{T}),$$

if for any $s, t \in \mathbb{T}$

$$C_1|t-s|^{2H} \leq \mathbb{E}(X_t - X_s)^2 \leq C_2|t-s|^{2H}.$$

Applying the independence of $B^{H_i}, 1 \leq i \leq N$, and formula [3.1] together with formula [3.4], we easily get the following statement concerning the incremental distance of the mfBm.

PROPOSITION 3.1.– *For all* $0 \leq s \leq t$,

$$E\Big(M_t^H(a) - M_s^H(a)\Big)^2 = \sum_{i=1}^{N} a_i^2 (t-s)^{2H_i}.$$

COROLLARY 3.2.– *For any* $T > 0$, *mfBm* $M^H(a)$ *belongs to the quasi-helix*

$$\mathbb{QH}(H_1, C_1, C_2, [0, T]),$$

with

$$C_1 = a_1^2 \text{ and } C_2 = N \max_{1 \leq k \leq N} \left(a_k^2 (2T)^{2H_k - 2H_1}\right).$$

Indeed,

$$\begin{aligned} a_1^2 (t-s)^{2H_1} &\leq \sum_{i=1}^{N} a_i^2 (t-s)^{2H_i} \\ &\leq \max_{1 \leq k \leq N} \left(a_k^2 (2T)^{2H_k - 2H_1}\right) (t-s)^{2H_1} \end{aligned}$$

Similarly, in the following proposition, we characterize the incremental distance of msfBm.

PROPOSITION 3.2.– *For all* $0 \leq s \leq t$

i) $$\mathbb{E}\left(S_t^H(a) - S_s^H(a)\right)^2 = \sum_{i=1}^{N} a_i^2 \left(-2^{2H_i-1}(t^{2H_i} + s^{2H_i}) + (t+s)^{2H_i} + (t-s)^{2H_i}\right). \quad [3.10]$$

ii) $$\sum_{i=1}^{N} a_i^2 \gamma_i (t-s)^{2H_i} \leq \mathbb{E}\Big(S_t^H(a) - S_s^H(a)\Big)^2 \leq \sum_{i=1}^{N} a_i^2 \nu_i (t-s)^{2H_i} \quad [3.11]$$

where

$$\gamma_i = \begin{cases} 2 - 2^{2H_i-1} \text{ if } H_i > 1/2, \\ 1 \text{ if } H_i \leq 1/2, \end{cases}$$

and

$$\nu_i = \begin{cases} 1 \text{ if } H_i > 1/2, \\ 2 - 2^{2H_i-1} \text{ if } H_i \leq 1/2. \end{cases}$$

PROOF.– Statement (i) is due to equality [2.11] and to the fact that the processes ξ^{H_i} are independent. Statement (ii) follows immediately from lemma 2.5. □

REMARK 3.3.– For any $T > 0$ and $s, t \in [0, T]$

$$\begin{aligned} a_1^2 \gamma_1 |t-s|^{2H_1} &\leq \sum_{i=1}^{N} a_i^2 \gamma_i (t-s)^{2H_i} \\ &\leq N \max_{1 \leq k \leq N} \left(a_k^2 \gamma_k (2T)^{2H_k - 2H_1}\right) |t-s|^{2H_1}. \end{aligned}$$

Therefore, for any $T > 0$, msfBm $S^H(a)$ belongs to the quasi-helix

$$\mathbb{QH}(H_1, C_1, C_2, [0, T]),$$

with

$$C_1 = a_1^2 \gamma_1, \; C_2 = N \max_{1 \leq k \leq N} \left(a_k^2 \gamma_k (2T)^{2H_k - 2H_1}\right).$$

REMARK 3.4.– On the one hand, it follows from proposition 3.1 that the mfBm has stationary increments. On the other hand, it follows from proposition 3.2 that the msfBm does not have stationary increments, but this property is replaced by inequalities [3.11]. Therefore, this is one of the main differences between these two processes.

In the following lemma, we present the explicit values of covariances of the increments of mfBm and msfBm on the non-overlapping intervals. The result is obtained by easy calculation, together with equalities [3.2] and [3.3]. For $0 \le u < v \le s < t$, introduce the following notations:

$$R(u,v,s,t) = R(u,v,s,t)(a) = \mathrm{Cov}\Big(M_v^H(a) - M_u^H(a), M_t^H(a) - M_s^H(a)\Big),$$

and

$$C(u,v,s,t) = C(u,v,s,t)(a) = \mathrm{Cov}\Big(S_v^H(a) - S_u^H(a), S_t^H(a) - S_s^H(a)\Big).$$

LEMMA 3.3.– *The incremental covariances of mfBm and msfBm equal*

$$R(u,v,s,t) = \sum_{i=1}^{N} \frac{a_i^2}{2}\left((t-u)^{2H_i} + (s-v)^{2H_i} - (t-v)^{2H_i} - (s-u)^{2H_i}\right),$$

and

$$C(u,v,s,t) = \sum_{i=1}^{N} \frac{a_i^2}{2}\left((t+u)^{2H_i} + (t-u)^{2H_i} + (s+v)^{2H_i} + (s-v)^{2H_i}\right.$$
$$\left. - (t+v)^{2H_i} - (t-v)^{2H_i} - (s+u)^{2H_i} - (s-u)^{2H_i}\right).$$

As the first consequence of lemma 3.3, we can specify the sign of correlation between the increments of our processes, according to the values of H.

COROLLARY 3.3.– *Let* $0 \le u < v \le s < t$.

i) If for any $1 \le i \le N$ *it holds that* $H_i \ge 1/2$ *and* $H_N > 1/2$, *then* $R(u,v,s,t) > 0$ *and* $C(u,v,s,t) > 0$.

ii) If for any $1 \le i \le N$ *it holds that* $H_i \le 1/2$ *and* $H_1 < 1/2$, *then* $R(u,v,s,t) < 0$ *and* $C(u,v,s,t) < 0$.

PROOF.– We will check the statement (i) for an smfBm, with other assertions being similar. Let us write

$$C(u,v,s,t) = \sum_{i=1}^{N} \frac{a_i^2}{2}(g_i(t) - g_i(s)),$$

where

$$g_i(t) = -(t+v)^{2H_i} - (t-v)^{2H_i} + (t+u)^{2H_i} + (t-u)^{2H_i}.$$

In the case under consideration, function g_i is differentiable for any $1 \le i \le N$, and for any $t > 0$

$$g_i'(t) = 2H_i\Big(- (t+v)^{2H_i-1} - (t-v)^{2H_i-1} + (t+u)^{2H_i-1} + (t-u)^{2H_i-1}\Big).$$

Moreover, $0 \le 2H_i - 1 \le 1$, and therefore the function $\psi : x \longmapsto x^{2H_i-1}$ is concave. In addition, we have that $t - v < t - u < t + u < t + v$. Hence, it follows from corollary A.1 that

$$\frac{\psi(t-u) - \psi(t-v)}{(t-u)-(t-v)} \ge \frac{\psi(t+v) - \psi(t+u)}{(t+v)-(t+u)},$$

which, in turn, implies that

$$\psi(t-u) + \psi(t+u) \ge \psi(t-v) + \psi(t+v).$$

Finally,

$$(t-u)^{2H_i-1} + (t+u)^{2H_i-1} \ge (t-v)^{2H_i-1} + (t+v)^{2H_i-1},$$

which yields $g_i'(t) > 0$. Consequently, g_i increases, and assertion (i) for an smfBm follows. □

As the second consequence of lemma 3.3, we can formulate the following result.

COROLLARY 3.4.–

i) Let all $H_i \ge 1/2$. *Then,* $C(u, v, s, t)(a_1, \dots, a_N)$ *increases in any* $|a_i|$. *Moreover, it increases strongly in* $|a_i|$ *if* $H_i > 1/2$.

ii) Let all $H_i \le 1/2$. *Then,* $C(u, v, s, t)(a_1, \dots, a_N)$ *decreases in any* $|a_i|$. *Moreover, it decreases strongly in* $|a_i|$ *if* $H_i > 1/2$.

In turn, this means that the dependence of increments in both cases increases with the absolute values of the coefficients. Consequently, in the modeling of a certain phenomenon, we can choose $H = (H_1, \dots, H_N)$ *and* $a = (a_1, \dots, a_N)$ *suitably in such a manner that* $\{S_t^H(a)\}$ *or* $\{M_t^H(a)\}$ *permits us to obtain a good model, taking into account not only the sign (as in the case of fractional Brownian motion and sub-fractional Brownian motion) but also the strength of dependence between the increments of the phenomenon.*

In the following lemma, we study the long- and short-range-dependent properties of the increments of our processes. In order to do this, we assume that x is a non-negative number and n is a non-negative integer. Denote

$$C(x,n) = C(x, x+1, x+n, x+n+1) = \mathrm{Cov}\Big(S_{x+1}^H - S_x^H, S_{x+n+1}^H - S_{x+n}^H\Big),$$

and

$$R(x,n) = R(x,x+1,x+n,x+n+1) = \mathrm{Cov}\Big(M^H_{x+1} - M^H_x, M^H_{x+n+1} - M^H_{x+n}\Big).$$

Let us first mention that the stationarity of the increments of M^H allows us to write the following equality

$$R(x,n) = R(0,n) = \mathrm{Cov}\Big(M^H_1, M^H_{n+1} - M^H_n\Big).$$

The next statement is the result of direct calculations.

LEMMA 3.4.– *For any $x \geq 0$ and $n \in \mathbb{N}$, we have*

i) $C(x,n) = \sum_{i=1}^N \frac{a_i^2}{2}\Big((n+1)^{2H_i} - 2n^{2H_i} + (n-1)^{2H_i}$

$$-(2x+n+2)^{2H_i} + 2(2x+n+1)^{2H_i} - (2x+n)^{2H_i}\Big).$$

ii) $$R(x,n) = R(0,n) = \sum_{i=1}^N \frac{a_i^2}{2}\Big((n+1)^{2H_i} - 2n^{2H_i} + (n-1)^{2H_i}\Big).$$

From lemma 3.4, we get the following statement.

COROLLARY 3.5.–

i) For any $x \geq 0$ and $H \in (0,1)^N$, we have that the following series converges: $\sum_{n \geq 0} C(x,n) < \infty$.

ii) Let $x \geq 0$. Then, $\sum_{n\geq 0} R(x,n) < \infty$ if and only if $H_i \leq \frac{1}{2}$ for any $1 \leq i \leq N$.

PROOF.– It follows from lemma 3.4 and the Taylor expansion that as $n \to \infty$, we have, similarly to lemma 2.8:

$$C(x,n) = \sum_{i=1}^N \Big(2(1-H_i)H_i(2H_i-1)(2x+1)a_i^2 n^{2H_i-3} + o(n^{2H_i-3})\Big)$$

and

$$R(x,n) = \sum_{i=1}^N \Big(H_i(2H_i-1)a_i^2 n^{2H_i-2} + o(n^{2H_i-2})\Big).$$

for any $H \in (0,1)^N$ and $x \geq 0$. From the above equalities, the stated corollary is easily obtained. □

REMARK 3.5.– Consequently, for any $H \in (0,1)^N$, the increments of S^H are short-range dependent, whereas M^H exhibits short-range dependence if and only if $H_i \leq \frac{1}{2}$ for any $1 \leq i \leq N$. Therefore, this is a second main difference between these two processes.

Now we shall analyze the function $x \longmapsto C(x,n)$ in order to understand "how far" the msfBm is from the mfBm. In other words, we want to understand "how far" the msfBm is from the class of processes with stationary increments.

LEMMA 3.5.– *For any $n \in \mathbb{N}$, it holds that*

$$C(x,n) = R(0,n) - \sum_{i=1}^{N} a_i^2 2^{2H_i-1} H_i(2H_i-1)x^{2(H_i-1)} + o(x^{2(H_i-1)}),$$

as $x \to \infty$. Therefore, $\lim_{x\to\infty} C(x,n) = R(0,n)$ for any $n \in \mathbb{N}$.

The result follows immediately from lemma 3.4 and standard Taylor expansion.

3.3. Invertibility of the covariance matrix of mfBm and msfBm

Consider any $n \in \mathbb{N}$ and, as usual, denote by v^T the transpose of a vector $v = (v_1, v_2, \ldots, v_n) \in \mathbb{R}^n$. Also, let $0 \leq t_1 \leq \cdots \leq t_n$ and $S_n^H = (S_{t_1}^H, \ldots, S_{t_n}^H)^T$.

In this section, we study the invertibility of the covariance matrix of the random vector S_n^H, and mfBm can be considered similarly. For this matrix, we use the notation

$$\Gamma = \left(\text{Cov}\left(S_{t_i}^H, S_{t_j}^H\right)\right)_{i,j=1,,\ldots,n}.$$

Let us start with the following important result.

THEOREM 3.2.– *Consider a sub-fractional Brownian motion ξ^H with parameter $H \in (0,1)$. Then, Gaussian random variables $\xi_{t_1}^H, \ldots, \xi_{t_n}^H$ are linearly independent.*

PROOF.– Let $\alpha_1, \ldots, \alpha_n \in \mathbb{R}$ be such that $\sum_{j=1}^{n} \alpha_j \xi_{t_j}^H = 0$ a.s. It follows from the spectral representation of the sfBm, contained in equality [2.14], that

$$\widetilde{C}(H) \sum_{j=1}^{n} \alpha_j \int_{\mathbb{R}} \frac{\cos(t_j x) - 1}{x} |x|^{\frac{1}{2}-H} dW(x) = 0 \; a.s. \qquad [3.12]$$

where the Wiener measure W together with the constant $\widetilde{C}(H)$ are described in theorem 2.2. It follows from equation [3.12] and the isometry property that

$$\mathbb{E}\left(\sum_{j=1}^{n}\alpha_j \int_{\mathbb{R}} \frac{\cos(t_j x)-1}{x}|x|^{\frac{1}{2}-H}dW(x)\right)^2$$

$$= \int_{\mathbb{R}} \left|\sum_{j=1}^{n}\alpha_j \frac{(\cos(t_j x)-1}{x}|x|^{\frac{1}{2}-H}\right|^2 dx = 0.$$

Thus, for almost any $x \in \mathbb{R}$, we have that

$$\left|\sum_{j=1}^{n}\alpha_j \frac{\cos(t_j x)-1}{x}|x|^{\frac{1}{2}-H}\right|^2 = 0,$$

which implies that

$$\sum_{j=1}^{n}\alpha_j \frac{\cos(t_j x)-1}{x}|x|^{\frac{1}{2}-H} = 0 \quad \text{a.e.}$$

In other words, we obtain that

$$\sum_{j=1}^{n}\alpha_j (\cos(t_j x)-1) = 0 \quad \text{a.e.}$$

Taking into account linear independence of the family of functions $\{f_1,\dots,f_n\}$, where $f_j : x \mapsto \cos(t_j x) - 1$ for any j, we obtain that $\alpha_j = 0$, for all $1 \leq j \leq n$, which permits us to conclude the proof. □

COROLLARY 3.6.– *The Gaussian random variables $S^H_{t_1},\dots,S^H_{t_n}$ are linearly independent.*

PROOF.– Let $\alpha_1,\dots,\alpha_n \in \mathbb{R}$ such that $\sum_{j=1}^{n}\alpha_j S^H_{t_j} = 0$ a.s. Then,

$$\mathbb{E}\left(\sum_{j=1}^{n}\alpha_j S^H_{t_j}\right)^2 = \mathbb{E}\left(\sum_{j=1}^{n}\alpha_j \sum_{k=1}^{N} a_k \xi^{H_k}_{t_j}\right)^2 = \mathbb{E}\left(\sum_{k=1}^{N} a_k \left(\sum_{j=1}^{n}\alpha_j \xi^{H_k}_{t_j}\right)\right)^2 = 0.$$

From the independence of the sub-fractional Brownian motions $\xi^{H_k}, 1 \leq k \leq N$, we get that

$$\sum_{k=1}^{N} a_k^2 \mathbb{E}\left(\sum_{j=1}^{n} \alpha_j \xi_{t_j}^{H_k}\right)^2 = 0.$$

This relation, together with our assumption $a_k \neq 0$, means that for any k

$$\mathbb{E}\left(\sum_{j=1}^{n} \alpha_j \xi_{t_j}^{H_k}\right)^2 = 0. \qquad [3.13]$$

Obviously, equation [3.13] means that

$$\sum_{j=1}^{n} \alpha_j \xi_{t_j}^{H_k} = 0$$

a.s. Now, it follows from the linear independence of $\xi_{t_1}^{H_k}, \ldots, \xi_{t_n}^{H_k}$, obtained in theorem 3.2, that $\alpha_j = 0$ for all $1 \leq j \leq n$, whence the proof follows. □

COROLLARY 3.7.– *The covariance matrix* Γ *of the random vector* $(S_{t_1}^H, \ldots, S_{t_n}^H)^T$ *is invertible.*

PROOF.– We know that the covariance matrix Γ is positive semi-definite. We will prove that it is also definite. Let $u = (u_1, \ldots, u_n)^T \in \mathbb{R}^n \setminus \{0\}$. We know that

$$u^T \Gamma u = \mathbb{E}\left(\sum_{j=1}^{n} u_j S_{t_j}^H\right)^2.$$

Let us suppose that $u^T \Gamma u = 0$. Then, $\mathbb{E}\left(\sum_{j=1}^{n} u_j S_{t_j}^H\right)^2 = 0$, which implies that

$$\sum_{j=1}^{n} u_j S_{t_j}^H = 0$$

a.s. From the linear independence of $S_{t_1}^H, \ldots, S_{t_n}^H$ established in corollary 3.6, we get that $u_j = 0$ for any j, which implies $u = 0$. This contradicts the fact that $u \in \mathbb{R}^n \setminus \{0\}$. Therefore, $u^T \Gamma u > 0$, for any $u \in \mathbb{R}^n \setminus \{0\}$, which means that Γ is definite, and consequently, it is invertible. □

3.4. Some properties of the mfBm's and msfBm's sample paths

Recall that the coefficients H_i are linearly ordered in increasing order and coefficients $a_i \neq 0$.

3.4.1. *Hölder continuity and non-differentiability of the sample paths of mfBm and msfBm*

In the following lemma, we check the Hölder continuity of the sample paths of the mixed fractional Gaussian processes and establish that the parameter H_1 controls their regularity.

LEMMA 3.6.– *For any $T > 0$ and $0 < \gamma < H_1$, each of the processes, mfBm and msfBm, admits a modification, whose sample paths are Hölder continuous of order γ on the interval $[0, T]$.*

PROOF.– We will check the lemma in the case of the msfBm. In the other case, the proof is similar. It follows from assertion (iii) in lemma 2.6 that for any $1 \leq i \leq N$ and $0 < \epsilon < H_i$, there exists a non-negative random variable $G_{H_i,\epsilon,T}$ such that $\mathbb{E}(G^p_{H_i,\epsilon,T}) < \infty$ for any $p \geq 1$, and

$$|\xi_t^{H_i} - \xi_s^{H_i}| \leq G_{H_i,\epsilon,T}|t - s|^{H_i - \epsilon} \quad a.s,$$

for all $s, t \in [0, T]$.

Therefore, for any $0 < \epsilon < H_1$, and $s, t \in [0, T]$

$$|S_t^H - S_s^H| = \left|\sum_{i=1}^{N} a_i(\xi_t^{H_i} - \xi_s^{H_i})\right| \leq G_{\epsilon,T}|t - s|^{H_1 - \epsilon} \quad a.s.$$

where

$$G_{\epsilon,T} = \sum_{i=1}^{N} |a_i| G_{H_i,\epsilon,T} T^{H_i - H_1}$$

for which we clearly have $\mathbb{E}(G^p_{\epsilon,T}) < \infty$ for any $p \geq 1$. Thus, the proof is finished. □

In the next theorem, we establish that the trajectories of the smfBm and also of the mfBm with probability 1 are not differentiable at any fixed point.

THEOREM 3.3.– *For any $H \in (0,1)^N$ and any $t_0 \in \mathbb{R}_+$, it holds that*

$$\lim_{\epsilon \downarrow 0} \sup_{t \in [(t_0-\epsilon)\wedge 0, t_0+\epsilon]} \left| \frac{S_t^H - S_{t_0}^H}{t - t_0} \right| = \lim_{\epsilon \downarrow 0} \sup_{t \in [(t_0-\epsilon)\wedge 0, t_0+\epsilon]} \left| \frac{M_t^H - M_{t_0}^H}{t - t_0} \right| = +\infty$$

with probability 1.

PROOF.– We will check the lemma in the case of msfBm. In the case of mfBm, the proof is even easier to achieve. Letting $t_0 > 0$, the case $t_0 = 0$ can be considered in a similar way. Let n_{t_0} be such integer that for $n \geq n_{t_0}$ $1/n < t_0$. For $m \in \mathbb{N}$, denote $A^{(m)} = \cap_{n=n_{t_0}}^{+\infty} A_n^{(m)}$, where, for $n \geq n_{t_0}$,

$$A_n^{(m)} = \left\{ \omega \in \Omega | \sup_{t \in [t_0 - \frac{1}{n}, t_0 + \frac{1}{n}]} \left| \frac{S_t^H - S_{t_0}^H}{t - t_0} \right| > m \right\}.$$

In order to obtain the result, it is enough to prove that $\mathbb{P}\Big(\cap_{m=1}^{+\infty} A^{(m)} \Big) = 1$.

On the one hand, the fact that each of the sequences $\left(A_n^{(m)} \right)_n$ and $\left(A^{(m)} \right)_m$ decreases allows us to write

$$\mathbb{P}\Big(\cap_{m=1}^{+\infty} A^{(m)} \Big) = \lim_{m \to +\infty} \mathbb{P}\Big(A^{(m)} \Big) = \lim_{m \to +\infty} \lim_{n \to +\infty} \mathbb{P}\Big(A_n^{(m)} \Big).$$

On the other hand,

$$\mathbb{P}(A_n^{(m)}) \geq \mathbb{P}\left(\left| S_{t_0+\frac{1}{n}}^H - S_{t_0}^H \right| > \frac{m}{n} \right).$$

Therefore, to prove the lemma, it is enough to establish that

$$\forall m \in \mathbb{N}, \lim_{n \to +\infty} \mathbb{P}\left(\left| S_{t_0+\frac{1}{n}}^H - S_{t_0}^H \right| \leq \frac{m}{n} \right) = 0.$$

Because $S_{t_0+\frac{1}{n}}^H - S_{t_0}^H$ is a centered Gaussian random variable with variance

$$\sigma_n^2(t_0) = \mathbb{E}\Big(S_{t_0+\frac{1}{n}}^H - S_{t_0}^H \Big)^2,$$

we have

$$\begin{aligned} \mathbb{P}\left(\left| S_{t_0+\frac{1}{n}}^H - S_{t_0}^H \right| \leq \frac{m}{n} \right) &= \frac{1}{\sigma_n(t_0)\sqrt{2\pi}} \int_{-\frac{m}{n}}^{\frac{m}{n}} \exp\left(-\frac{x^2}{2\sigma_n^2(t_0)} \right) dx \\ &\leq 2\frac{m}{n} \cdot \frac{1}{\sigma_n(t_0)\sqrt{2\pi}}. \end{aligned}$$

It follows from proposition 3.2 that

$$\sigma_n^2(t_0) = \sum_{i=1}^{N} a_i^2 \Bigg(-2^{2H_i-1} \left(\left(t_0 + \frac{1}{n}\right)^{2H_i} + t_0^{2H_i} \right) + \left(2t_0 + \frac{1}{n}\right)^{2H_i} + \frac{1}{n^{2H_i}} \Bigg).$$

Therefore, for large n, we have that $\sigma_n^2(t_0) \approx \sum_{i=1}^{N} a_i^2 n^{-2H_i}$.

Therefore, $\lim_{n\to+\infty} n^2\sigma_n^2(t_0) = \lim_{n\to+\infty} \sum_{i=1}^{N} a_i^2 n^{2-2H_i} = +\infty$, and consequently

$$\lim_{n\to+\infty} \mathbb{P}\left(\left| S_{t_0+\frac{1}{n}}^H - S_{t_0}^H \right| \leq \frac{m}{n} \right) = 0. \qquad \square$$

REMARK 3.6.– There is a much stronger result: the trajectories of the mixed processes with probability 1 are not differentiable at all points of any interval. However, the proof of this fact is more complicated. For the Wiener process, it is a well-known Paley–Wiener–Zygmund–Dvoretzky–Erdös–Kakutani theorem, see, e.g. [MIS 17], theorem 4.3, and the references therein.

3.4.2. *Hausdorff dimensions of some sets related to the sample paths of mfBm and msfBm*

Introduce the following notations. Namely, denote the range of the restriction of $S^H(a)$ on $[0,T]$ by

$$S^H([0,T]) = \left\{ S_t^H | t \in [0,T] \right\}, \qquad [3.14]$$

its graph by

$$\mathrm{Grf}_T S^H(a) = \left\{ (t, S_t^H(a)) | t \in [0,T] \right\}, \qquad [3.15]$$

the graph of $S^H(a)$ by

$$\mathrm{Grf} S^H(a) = \left\{ (t, S_t^H(a)) | t \in [0,+\infty) \right\}, \qquad [3.16]$$

and the level set of the restriction of $S^H(a)$ on $[\epsilon, T]$ by

$$L_x^\epsilon = \Big\{t \in [\epsilon, T] | S_t^H = x\Big\}, \qquad [3.17]$$

where $0 < \epsilon < T$. Now, our goal is to study the Hausdorff dimensions of the sets defined in equations [3.14]–[3.17]. A similar study can be provided for the process $M^H(a)$. We leave it for the reader as an exercise. A brief description of the definition of the Hausdordff dimension and its properties is presented in section A.5 of the Appendix.

In the following, we need the following auxiliary lemma.

LEMMA 3.7.– *For all $(s,t) \in \mathbb{R}_+^2$ such that $s \neq t$, and for any real $u > 1$, we have that*

$$\mathbb{E}\left(\Big(|s-t| + |S_t^H(a) - S_s^H(a)|\Big)^{-u}\right) \leq c|t-s|^{1-u}\sigma^{-1}(s,t), \qquad [3.18]$$

where

$$\begin{aligned}\sigma^2(s,t) &= \mathbb{E}\Big(S_t^H(a) - S_s^H(a)\Big)^2 \\ &= \sum_{i=1}^N a_i^2\left(-2^{2H_i-1}(t^{2H_i} + s^{2H_i}) + (t+s)^{2H_i} + (t-s)^{2H_i}\right)\end{aligned}$$

and $c = c(u) > 0$ is a constant not depending on s, t and H.

PROOF.– Let us rewrite and bound from above the left-hand side of equation [3.18]:

$$\begin{aligned}&\mathbb{E}\left(\Big(|s-t| + |S_t^H(a) - S_s^H(a)|\Big)^{-u}\right) \\ &= \frac{1}{\sigma(s,t)\sqrt{2\pi}}\int_{\mathbb{R}}\Big(|t-s| + |x|\Big)^{-u}\exp\left(-\frac{x^2}{2\sigma^2(s,t)}\right)dx \\ &= \frac{2}{\sigma(s,t)\sqrt{2\pi}}\int_0^{|t-s|}\Big(|t-s| + |x|\Big)^{-u}\exp\left(-\frac{x^2}{2\sigma^2(s,t)}\right)dx \\ &+ \frac{2}{\sigma(s,t)\sqrt{2\pi}}\int_{|t-s|}^{+\infty}\Big(|t-s| + |x|\Big)^{-u}\exp\left(-\frac{x^2}{2\sigma^2(s,t)}\right)dx \\ &\leq \frac{2}{\sigma(s,t)\sqrt{2\pi}}\left(\int_0^{|t-s|}|t-s|^{-u}dx + \int_{|t-s|}^{+\infty}|x|^{-u}dx\right) \\ &\leq \frac{2}{\sigma(s,t)\sqrt{2\pi}}\left(|t-s|^{1-u} + \frac{1}{u-1}|t-s|^{1-u}\right) \leq c|t-s|^{1-u}\sigma^{-1}(s,t),\end{aligned}$$

with

$$c = \frac{\sqrt{2}}{\sqrt{\pi}} \frac{u}{u-1},$$

and lemma is proved. □

The following theorem characterizes the Hausdorff dimension of the range $S^H([0,T])$.

THEOREM 3.4.– *The Hausdorff dimension of the range $S^H([0,T])$ equals 1 with probability 1.*

PROOF.– It follows from lemma 3.6 that

$$\mathbb{P}\left(\omega \in \Omega | t \mapsto S_t^H \text{ is continuous on } [0,T]\right) = 1.$$

Thus,

$$\mathbb{P}\left(\omega \in \Omega | S^H([0,T]) = [a_\omega, b_\omega],\ (a_\omega, b_\omega) \in \mathbb{R}^2, a_\omega < b_\omega\right) = 1.$$

By using corollary A.1 in the Appendix, $\dim[a,b] = 1$, for any real $a < b$. Consequently,

$$\mathbb{P}\Big\{\omega \in \Omega | \dim S^H([0,T]) = 1\Big\} = 1.$$ □

Now, the following theorem deals with the Hausdorff dimension of the graph of $S^H(a)$.

THEOREM 3.5.– *The Hausdorff dimension of the graph of $S^H(a)$ equals $2 - H_1$ with probability* 1.

PROOF.– It was established in lemma 3.6 that for all $T > 0$ the msfBm has a modification, whose sample paths are Hölder continuous of order $\gamma < H_1$ on the interval $[0,T]$. Therefore, according to lemma A.10, for all $T > 0$, with probability 1,

$$\dim \mathrm{Grf}_T S^H(a) \leq 2 - H_1,$$

which implies, by using the third assertion in lemma A.1 in the Appendix, that

$$\dim \mathrm{Grf} S^H(a) \leq 2 - H_1.$$

To get the lower bound, we will first prove that

$$\int_0^T \int_0^T \left(|s-t| + |S_s^H(a) - S_t^H(a)|\right)^{-u} dsdt < +\infty \qquad [3.19]$$

a.s. for any $u \in (1, 2 - H_1)$.

It follows from lemma 3.7, together with proposition 3.2, that

$$\int_0^T \int_0^T \mathbb{E}\left(\left(|s-t| + |S_s^H(a) - S_t^H(a)|\right)^{-u}\right) dsdt$$
$$\leq \int_0^T \int_0^T c|t-s|^{1-u} \sigma^{-1}(s,t) dsdt \leq \int_0^T \int_0^T \frac{c}{a_1\sqrt{\gamma_1}} |t-s|^{1-u-H_1} dsdt,$$

where γ_1 is defined by equation [3.11]. Because $u \in \Big(1, 2 - H_1\Big)$, it is easy to check that

$$\int_0^T \int_0^T |t-s|^{1-u-H_1} dsdt < +\infty,$$

which shows that

$$\mathbb{E}\left(\int_0^T \int_0^T \left(|s-t| + |S_s^H(a) - S_t^H(a)|\right)^{-u} dsdt\right)$$
$$= \int_0^T \int_0^T \mathbb{E}\left(\left(|s-t| + |S_s^H(a) - S_t^H(a)|\right)^{-u}\right) dsdt < +\infty, \qquad [3.20]$$

and in turn it allows us to deduce equation [3.19].

To continue the proof, introduce an occupation measure $\nu(A)$. For any Borel set $A \subset \mathbb{R}^2$, $\nu(A)$ is defined as the integral

$$\nu(A) = \lambda_1\{t \in [0,T] | (t, S_t^H(a)) \in A\} = \int_0^T \mathbf{1}_{\{(t, S_t^H(a)) \in A\}} dt,$$

where λ_1 denotes the Lebesgue measure on $[0,T]$ and $\mathbf{1}_U$ denotes the characteristic function of the set U for any set $U \subset \mathbb{R}^2$.

Now, due to corollary A.5 in the Appendix, in order to obtain the lower bound, $\dim \mathrm{Grf}_T S^H(a) \leq 2 - H_1$, we only need to establish that for any $T > 0$, the occupation measure ν of $t \longmapsto (t, S_t^H(a))$, when t is restricted to the interval $[0,T]$ with probability 1, has a finite u-dimensional energy, for any $u \in \Big(1, 2 - H_1\Big)$.

To start with, we note that the measure ν is clearly carried by $\mathrm{Grf}_T S^H(a)$ and, denoting by $\|(x_1, x_2)\|_1 = |x_1| + |x_2|$ for any $(x_1, x_2) \in \mathbb{R}^2$, we obtain that

$$\mathbb{E}\left(\int_{\mathrm{Grf}_T S^H(a)} \int_{\mathrm{Grf}_T S^H(a)} \|x-y\|_1^{-u} \nu(dx)\nu(dy)\right)$$
$$= \mathbb{E}\left(\int_0^T \int_0^T \left(|s-t| + |S_s^H(a) - S_t^H(a)|\right)^{-u} dsdt\right)$$

which is finite by virtue of equation [3.20].

Consequently, the integral

$$\int_{\mathrm{Grf}_T S^H(a)} \int_{\mathrm{Grf}_T S^H(a)} \|x-y\|_1^{-u} \nu(dx)\nu(dy) \qquad [3.21]$$

is finite with probability 1. Now, note that the norms $\|\cdot\|_1$ and standard norm $|x| = (\sum_{i=1}^2 x_i^2)^{1/2}$ are equivalent; therefore, we conclude from [3.21] that

$$\int_{\mathrm{Grf}_T S^H(a)} \int_{\mathrm{Grf}_T S^H(a)} |x-y|_1^{-u} \nu(dx)\nu(dy)$$

is finite, and the proof follows from corollary A.5 in the Appendix. □

The following lemma is necessary for the study of the Hausdorff dimension of the level set L_x^ϵ. Recall that $\mathrm{Var}(Y|X)$ stands for the conditional variance of Y given X.

LEMMA 3.8.– *There exists a constant $c > 0$ such that for all $s, t \in [0, T]$,*

$$\mathrm{Var}(S_t^H | S_s^H) \geq c|s-t|^{2H_1}. \qquad [3.22]$$

PROOF.– Let $t > s$. Because S^H is a centered Gaussian process, it follows from corollary 1.3 that

$$\mathrm{Var}(S_t^H | S_s^H) = \inf_{b \in \mathbb{R}} \mathbb{E}\left(S_t^H - bS_s^H\right)^2.$$

Therefore, we can deduce from definition 3.1 of the msfBm, independence of the sfBm's ξ^{H_i} and corollary 1.3 that

$$\begin{aligned}\mathrm{Var}(S_t^H | S_s^H) &= \inf_{b\in\mathbb{R}} \sum_{i=1}^{N} \mathbb{E}\left(\xi_t^{H_i} - b\xi_s^{H_i}\right)^2 \\ &\geq \inf_{b\in\mathbb{R}} \mathbb{E}\left(\xi_t^{H_1} - b\xi_s^{H_1}\right)^2 = \mathrm{Var}(\xi_t^{H_1}|\xi_s^{H_1}) \\ &= \mathbb{E}(\xi_t^{H_1})^2 - \frac{\mathrm{Cov}^2(\xi_t^{H_1}, \xi_s^{H_1})}{\mathbb{E}(\xi_s^{H_1})^2} = \left(2 - 2^{2H_1-1}\right) t^{2H_1} \qquad [3.23] \\ &\quad - \frac{\left(t^{2H_1} + s^{2H_1} - 1/2(t+s)^{2H_1} - 1/2(t-s)^{2H_1}\right)^2}{\left(2 - 2^{2H_1-1}\right) s^{2H_1}}.\end{aligned}$$

Denote $s = \alpha t, \alpha \in [0,1]$. Then,

$$\begin{aligned}&\left(2 - 2^{2H_1-1}\right) t^{2H_1} - \frac{\left(t^{2H_1} + s^{2H_1} - 1/2(t+s)^{2H_1} - 1/2(t-s)^{2H_1}\right)^2}{\left(2 - 2^{2H_1-1}\right) s^{2H_1}} \\ &= \left(2 - 2^{2H_1-1}\right) t^{2H_1} \\ &\quad - \frac{t^{4H_1}\left(1 + (st^{-1})^{2H_1} - 1/2(1+st^{-1})^{2H_1} - 1/2(1-(st^{-1}))^{2H_1}\right)^2}{\left(2 - 2^{2H_1-1}\right) s^{2H_1}} \\ &= \frac{t^{2H_1}}{2 - 2^{2H_1-1}} \left(\left(2 - 2^{2H_1-1}\right)^2 - \alpha^{-2H_1}\left(1 + \alpha^{2H_1} - 1/2(1+\alpha)^{2H_1}\right.\right. \\ &\quad \left.\left. -1/2(1-\alpha)^{2H_1}\right)^2 \right) = \frac{(t-s)^{2H_1}}{2 - 2^{2H_1-1}} \left(\left(2 - 2^{2H_1-1}\right)^2 \right. \qquad [3.24] \\ &\quad \left. - \alpha^{-2H_1}\left(1 + \alpha^{2H_1} - 1/2(1+\alpha)^{2H_1} - 1/2(1-\alpha)^{2H_1}\right)^2 \right)(1-\alpha)^{-2H_1}.\end{aligned}$$

Consider the function

$$f(\alpha) = \frac{\left(2 - 2^{2H_1-1}\right)^2 - \alpha^{-2H_1}\left(1 + \alpha^{2H_1} - 1/2(1+\alpha)^{2H_1} - 1/2(1-\alpha)^{2H_1}\right)^2}{(1-\alpha)^{2H_1}}.$$

It is continuous on $(0,1)$ and positive for any $\alpha \in (0,1)$; therefore, it is necessary to consider its limit at points $\alpha = 0$ and $\alpha = 1$.

We have that

$$\lim_{\alpha\to 0} f(\alpha) = \lim_{\alpha\to 0}\left(\left(2-2^{2H_1-1}\right)^2 - \alpha^{-2H_1}\left(1+\alpha^{2H_1}\right.\right.$$
$$\left.\left. -1/2(1+\alpha)^{2H_1} - 1/2(1-\alpha)^{2H_1}\right)^2 \right)(1-\alpha)^{-2H_1}$$
$$= \lim_{\alpha\to 0}\frac{\left(2-2^{2H_1-1}\right)^2 - \alpha^{2H_1}}{(1-\alpha)^{2H_1}}$$
$$= \left(2-2^{2H_1-1}\right)^2 > 0.$$

Now, assuming that $\alpha \to 1$, we get that

$$1+\alpha^{2H_1} - 1/2(1+\alpha)^{2H_1} - 1/2(1-\alpha)^{2H_1} \sim 2 - 2^{2H_1-1} - 1/2(1-\alpha)^{2H_1}.$$

Thus,

$$f(\alpha) \sim \frac{\left(2-2^{2H_1-1}\right)^2 - \left(2-2^{2H_1-1} - 1/2(1-\alpha)^{2H_1}\right)^2}{(1-\alpha)^{2H_1}}$$
$$\sim 2 - 2^{2H_1-1} - \frac{1}{4}(1-\alpha)^{2H_1}.$$

Therefore, $\lim_{\alpha\to 1} f(\alpha) = 2 - 2^{2H_1-1} > 0$.

Consequently, f admits a continuous extension $\tilde{f}$ on $[0,1]$, and there exists $c > 0$ such that $\tilde{f}(\alpha) \geq c$, for any $\alpha \in [0,1]$.

Combining [3.23], [3.24] and the properties of extension $\tilde{f}$ considered above, we get that

$$\mathrm{Var}(\xi_t^{H_1} | \xi_s^{H_1}) \geq c|t-s|^{2H_1}, \qquad [3.25]$$

and equation [3.23], together with equation [3.25], completes the proof. □

REMARK 3.7.– According to theorem 2.1 from [YAN 10], for any $0 < a < b$, there exist positive constants c_1 and r_0 such that for all $t \in [a,b]$ and all $0 < r \leq \min(t, r_0)$,

$$\mathrm{Var}(\xi_t^{H_1} | \xi_s^{H_1},\ s \in [a,b], r \leq |s-t| \leq r_0) \geq c_1 r^{2H_1}.$$

Of course, inequality [3.25] is a direct consequence of this statement; however, we preferred to give its direct proof.

THEOREM 3.6.– *For any* $0 < \epsilon < T$*, with probability 1, a sample function* $S^H : [\epsilon, T] \mapsto \mathbb{R}$ *satisfies* $\dim L_x^\epsilon \leq 1 - H_1$ *for almost all* x *with respect to the Lebesgue measure on* $\mathbb{R}$*. Moreover, for any* $x \in \mathbb{R}$*, with positive probability*

$$\dim(L_x^\epsilon) = 1 - H_1. \tag{3.26}$$

PROOF.– For an integer $n \geq 1$, divide the interval $[\epsilon, T]$ into m_n sub-intervals $I_{n,l}$ of length n^{-1/H_1}. Then,

$$m_n \leq T n^{1/H_1}. \tag{3.27}$$

Let $0 < \delta < 1$ be fixed and let $\tau_{n,l} = \epsilon + n^{-1/H_1} l$. Then,

$$\begin{aligned} \mathbb{P}\left(x \in S^H(I_{n,l})\right) &\leq \mathbb{P}\left(\max_{s,t \in I_{n,l}} |S_s^H - S_t^H| \leq n^{-(1-\delta)}; x \in S^H(I_{n,l})\right) \\ &\quad + \mathbb{P}\left(\max_{s,t \in I_{n,l}} |S_s^H - S_t^H| > n^{-(1-\delta)}\right) \qquad [3.28] \\ &\leq \mathbb{P}\left(|S^H(\tau_{n,l}) - x| \leq n^{-(1-\delta)}\right) + \mathbb{P}\left(\max_{s,t \in I_{n,l}} |S_s^H - S_t^H| > n^{-(1-\delta)}\right). \end{aligned}$$

Let us evaluate both terms on the right-hand side of equation [3.28]. On the one hand, we have

$$\mathbb{P}\left(|S^H(\tau_{n,l}) - x| \leq n^{-(1-\delta)}\right) = \int_{-n^{-(1-\delta)}}^{n^{-(1-\delta)}} \frac{1}{\sqrt{2\pi\sigma^2(\tau_{n,l})}} \exp\left(-\frac{(u-m)^2}{2\sigma^2(\tau_{n,l})}\right) du,$$

where $\sigma^2(\tau_{n,l}) = \mathbb{E}\left(S^H(\tau_{n,l}) - x\right)^2$ and $m = \mathbb{E}\left(S^H(\tau_{n,l}) - x\right) = -x$. It follows from inequalities [3.11] established in proposition 3.2 that

$$\begin{aligned} \mathbb{E}\left(S^H(\tau_{n,l}) - x\right)^2 &\geq \sum_{i=1}^N a_i^2 \gamma_i (\tau_{n,l})^{2H_i} \\ &\geq a_1^2 \gamma_1 (\tau_{n,l})^{2H_1} = C(\epsilon + n^{-1/H_1} l)^{2H_1} \geq C\epsilon^{2H_1} \end{aligned}$$

where C denotes a generic constant. Thus, we get

$$\mathbb{P}\left(|S^H(\tau_{n,l}) - x| \leq n^{-(1-\delta)}\right) \leq 2n^{-(1-\delta)} \frac{1}{\sqrt{2\pi\sigma^2(\tau_{n,l})}} \leq C n^{-1+\delta}.$$

On the other hand, to evaluate $\mathbb{P}\left(\max_{s,t\in I_{n,l}} |S_s^H - S_t^H| > n^{-(1-\delta)}\right)$, we write that

$$\mathbb{P}\left(\max_{s,t\in I_{n,l}} |S_s^H - S_t^H| > n^{-(1-\delta)}\right) \le \mathbb{P}\left(\max_{t\in I_{n,l}} |S_t^H - S^H(\tau_{n,l})| > 1/2n^{-(1-\delta)}\right).$$

Now, consider a Gaussian process $X_t = S_t^H - S^H(\tau_{n,l})$ and apply theorem 1.3 to it with

$$\rho(\varepsilon) \le C\varepsilon^{H_1},\ Q(\delta) \le C\delta^{H_1},\ Q^{-1}(x) \ge Cx^{1/H_1}\ Q^{-1}(1/x) \ge Cx^{-1/H_1},$$

getting that

$$\begin{aligned}&\mathbb{P}\left(\max_{t\in I_{n,l}} |S_t^H - S^H(\tau_{n,l})| > 1/2n^{-(1-\delta)}\right)\\ &\le C\left(n^{-(1-\delta)}\right)^{1/H_1} \le Cn^{-(1-\delta)}.\end{aligned} \qquad [3.29]$$

Summarizing equations [3.28]–[3.29], we get the upper bound

$$\mathbb{P}\left(x \in S^H(I_{n,l})\right) \le Cn^{-(1-\delta)}. \qquad [3.30]$$

Now, define a covering $\{I'_{n,l}\}$ of L_x by $I'_{n,l} = I_{n,l}$ if $x \in S^H(I_{n,l})$ and $I'_{n,l} = \emptyset$ otherwise. Denote the number of non-empty sets $\{I'_{n,l}\}$ by M_n. By using equation [3.27] and equation [3.30], we get that

$$\begin{aligned}\mathbb{E}(M_n) &\le \mathbb{E}\sum_{l=0}^{m_n}\left(\mathbf{1}_{\{x\in S^H(I_{n,l})\}}\right) = \sum_{l=0}^{m_n}\mathbb{P}\left(x \in S^H(I_{n,l})\right)\\ &\le CTn^{1/H_1}n^{-1+\delta} \le C_6 n^{1/H_1-1+\delta},\end{aligned} \qquad [3.31]$$

where C_6 is a positive constant.

Let $\eta = 1/H_1 - (1 - 2\delta)$. From equation [3.31] and the Markov inequality, we get that almost surely $M_n \le C_7 n^\eta$ for all n large enough. This implies that the $H_1\eta$-Hausdorff measure of L_x^ϵ that is denoted by $M^{H_1\eta}(L_x^\epsilon)$, is finite. Indeed,

$$\begin{aligned}M^{H_1\eta}(L_x^\epsilon) &\le \lim_{n\to\infty}\inf\left\{\sum_{k=1}^{M_n}|I'_{n,k}|^{H_1\eta};\ \{I'_{n,k}\}_{k=1}^{M_n} \text{ is a } n^{-1/H_1} - \text{covering of } L_x^\epsilon\right\}\\ &\le \lim_{n\to\infty} CM_n(n^{-1/H_1})^{H_1\eta} \le \lim_{n\to\infty} Cn^\eta n^{-\eta} = C < \infty.\end{aligned}$$

Consequently, $\dim L_x^\epsilon \leq H_1\eta$ almost surely. Letting $\delta \downarrow 0$ along rational numbers, we get

$$\dim L_x^\epsilon \leq 1 - H_1 \quad a.s.$$

To prove the lower bound for $\dim L_x^\epsilon$ in equation [3.26], we consider a small constant $\delta > 0$ such that

$$\gamma = 1 - H_1(1+\delta) > 0.$$

Note that if we can prove that there is a constant $C_8 > 0$, independent of δ and such that

$$\mathbb{P}\Big\{ \dim L_x^\epsilon \geq \gamma \Big\} \geq C_8, \qquad [3.32]$$

then the lower bound in equation [3.26] will follow by letting $\delta \downarrow 0$.

Our proof of equation [3.32] is based on the capacity argument contained in Kahane's book (see [KAH 85]). Similar methods have been used in [ADL 81, AYA 05, TES 87, XIA 08].

Consider the space $\mathcal{M}_\gamma^+$ of all non-negative measures on $\mathbb{R}$ with finite γ-energy (see definition A.16). By using corollary A.4 and theorem A.4, $\mathcal{M}_\gamma^+$ is a complete metric space under the metric defined by

$$d : (\mu, \nu) \longmapsto \int_{\mathbb{R}} \int_{\mathbb{R}} \frac{(\mu-\nu)(dt)(\mu-\nu)(ds)}{|t-s|^\gamma}.$$

We define a sequence of random positive measures $\mu_n := \mu_n(C, \omega)$ on the Borel sets C of $[\epsilon, T]$ by

$$\begin{aligned} \mu_n(C) = \mu_n(C,\omega) &= \int_C \sqrt{2n\pi} \exp\Big(- \frac{n(S^H(t)-x)^2}{2} \Big) dt \\ &= \int_C \int_{\mathbb{R}} \exp\Big(- \frac{\xi^2}{2n} + i\xi(S^H(t)-x) \Big) d\xi dt \qquad [3.33] \\ &= \int_C \Phi_n(t,\omega) dt \end{aligned}$$

where for each $n \geq 1$,

$$\Phi_n(t,\omega) = \int_{\mathbb{R}} \exp\Big(- \frac{\xi^2}{2n} + i\xi(S^H(t)-x) \Big) d\xi.$$

As was shown in [FEL 71], Chapter 15, the function $\Phi_n(t) \geq 0$ is a continuous function, and with probability 1 $\Phi_n(t)$ weakly converges to zero outside any neighborhood of $(S^H)^{-1}\{x\}$ as $n \to \infty$. That is, with probability 1,

$$\lim_{n\to\infty} \int_C \Phi_n(t,\omega)g(t,\omega)dt = 0,$$

for any continuous function g that vanishes in a neighborhood of $(S^H)^{-1}\{x\}$.

It follows from [KAH 85] (p. 206) or [TES 87] (p.17) that if there exist positive and finite constants C_9, C_{10} and C_{11} such that

$$\mathbb{E}(\|\mu_n\|) \geq C_9, \quad \mathbb{E}(\|\mu_n\|^2) \leq C_{10}, \qquad [3.34]$$

and

$$\mathbb{E}(\|\mu_n\|_\gamma) \leq C_{11}, \qquad [3.35]$$

where $\|\mu_n\| = \mu_n\left([\epsilon, T]\right)$ denotes the total mass of μ_n, and

$$\|\mu_n\|_\gamma = \int_{\mathbb{R}}\int_{\mathbb{R}} \frac{\mu_n(dt)\ \mu_n(ds)}{|t-s|^\gamma},$$

then there is a subsequence of $\{\mu_n\}$, say $\{\mu_{n_k}\}$, such that $\mu_{n_k} \to \mu$ in $\mathcal{M}_\gamma^+$ and μ is strictly positive with probability $\geq C_9^2/(2C_{10})$. In this case, it follows from equation [3.33] that μ has its support in L_x^ϵ almost surely. Moreover, equation [3.35] implies that the γ-energy of μ is finite. Hence, Frostman's theorem yields equation [3.32] with $C_{12} = C_9^2/(2C_{10})$.

It remains to verify equations [3.34] and [3.35]. By using Fubini's theorem, we get

$$\begin{aligned}
\mathbb{E}(\|\mu_n\|) &= \int_\epsilon^T \int_{\mathbb{R}} \exp(-i\xi x) \exp\Big(-\frac{\xi^2}{2n}\Big)\mathbb{E}\Big(\exp(i\xi S^H(t))\Big) d\xi dt \\
&= \int_\epsilon^T \int_{\mathbb{R}} \exp(-i\xi x) \exp\Big(-\frac{1}{2}(n^{-1}+\sigma^2(t))\xi^2\Big) d\xi dt \\
&= \int_\epsilon^T \sqrt{\frac{2\pi}{n^{-1}+\sigma^2(t)}} \exp\Big(-\frac{x^2}{2(n^{-1}+\sigma^2(t))}\Big) dt \\
&\geq \int_\epsilon^T \sqrt{\frac{2\pi}{1+\sigma^2(t)}} \exp\Big(-\frac{x^2}{2\sigma^2(t)}\Big) dt := C_9.
\end{aligned}$$

Denote by I_2 the identity matrix of order 2 and by $\text{Cov}(S_s^H, S_t^H)$ the covariance matrix of the Gaussian vector (S_s^H, S_t^H). Let $\Gamma = n^{-1}I_2 + \text{Cov}(S_s^H, S_t^H)$ and $(\xi, \eta)^T$ be the transpose of the row vector (ξ, η). Then,

$$\begin{aligned}
\mathbb{E}(\|\mu_n\|^2) &= \int_\epsilon^T \int_{\mathbb{R}} \int_\epsilon^T \int_{\mathbb{R}} \exp(-i(\xi+\eta)x) \exp\Big(-\frac{\xi^2+\eta^2}{2n}\Big) \\
&\quad \times \mathbb{E}\Big(\exp(i(\xi S_t^H + \eta S_s^H))\Big) d\xi d\eta dt ds \\
&= \int_\epsilon^T \int_\epsilon^T \int_{\mathbb{R}} \int_{\mathbb{R}} \exp\Big(-i(\xi+\eta)x\Big) \\
&\quad \exp\Big(-(\xi,\eta)\Gamma(\xi,\eta)^T/2\Big) d\xi d\eta ds dt \qquad [3.36] \\
&= 2\pi \int_\epsilon^T \int_\epsilon^T (\det \Gamma)^{-1/2} \exp\Big(-\frac{1}{2}(x,x)\Gamma^{-1}(x,x)^T\Big) ds dt \\
&\leq 2\pi \int_\epsilon^T \int_\epsilon^T (\det \text{Cov}(S_s^H, S_t^H))^{-1/2} ds dt.
\end{aligned}$$

Note that

$$\det \text{Cov}(S_s^H, S_t^H) = \text{Var}(S_s^H)\text{Var}\Big(S_t^H | S_s^H\Big). \qquad [3.37]$$

Therefore, it follows from equality [3.5] and inequality [3.22] that for all $s, t \in [\epsilon, T]$,

$$\det \text{Cov}(S_s^H, S_t^H) \geq C_{13}|s-t|^{2H_1}. \qquad [3.38]$$

where C_{13} denotes a positive constant. Combining equations [3.37] and [3.38], we obtain that

$$\mathbb{E}(\|\mu_n\|^2) \leq C_{14} \int_\epsilon^T \int_\epsilon^T |s-t|^{-H_1} ds dt.$$

where C_{14} denotes a positive constant. Because $H_1 \in (0,1)$, the last integral is a finite constant, and consequently

$$\mathbb{E}(\|\mu_n\|^2) \leq C_{10}$$

with $C_{10} = C_{14} \int_\epsilon^T \int_\epsilon^T |s-t|^{-H_1} ds dt.$

Similarly to equation [3.36], we have

$$\begin{aligned}\mathbb{E}(\|\mu_n\|_\gamma) &= \int_\epsilon^T \int_\epsilon^T |s-t|^{-\gamma} dsdt \\ &\times \int_{\mathbb{R}} \int_{\mathbb{R}} \exp\Big(-i(\xi+\eta)x\Big) \exp\Big(-(\xi,\eta)\Gamma(\xi,\eta)^T/2\Big) d\xi d\eta \\ &\le C_{15} \int_\epsilon^T \int_\epsilon^T |s-t|^{-\gamma-H_1} dsdt,\end{aligned}$$

where C_{15} denotes a positive constant. Because $-\gamma - H_1 = -1 + H_1\delta \in (-1, 0)$, we get that $\mathbb{E}(\|\mu_n\|_\gamma) < C_{11}$ with

$$C_{11} = C_{15} \int_\epsilon^T \int_\epsilon^T |s-t|^{-\gamma-H_1} dsdt,$$

and the proof is completed. □

3.5. A series expansion of mixed sub-fractional Brownian motion

In this section, we focus our attention on the particular interesting case where $N = 2$ and $H_1 = 1/2$. We present an explicit series expansion of the mixed sub-fractional Brownian motion and study the rate of its convergence. As an application of our result, we present a computer simulation of mixed sub-fractional Brownian motion sample paths.

3.5.1. *Explicit series expansion of mixed sub-fractional Brownian motion*

Recall that for $\nu \neq -1, -2, \ldots$ the Bessel function J_ν of the first kind of order ν can be defined on the set $\{z \in \mathbb{C} | \, |arg(z)| < \pi\}$ as the absolutely convergent series

$$J_\nu(z) = \sum_{n=0}^{\infty} \frac{(-1)^n (z/2)^{\nu+2n}}{\Gamma(n+1)\Gamma(\nu+n+1)},$$

where $\Gamma(x)$ is, as always, Euler's gamma function. It is well known that for $\nu > -1$, the function J_ν has countable number of real, positive, simple zeros (see [WAT 44], Chapter 15). These zeros $x_1 < x_2 < \cdots < x_n < \cdots$ can be arranged in ascending order of magnitude and $x_n \to \infty$ as $n \to \infty$. The next lemma follows from the asymptotic properties of the Bessel function and its positive zeros (see [WAT 44]).

LEMMA 3.9.– *For $\nu > -1$, let J_ν be the Bessel function of the first kind of order νá and let $z_1 < z_2 < \cdots$ be its positive zeros. Then,* *i)* $z_n \sim n\pi$, *and* $J^2_{\nu+1}(z_n) \sim 2/n\pi^2$ *as* $n \to \infty$,

ii) $J_{\nu-1}(z_n) + J_{\nu+1}(z_n) = 0$, *for any* $n \geq 1$.

Now, let $H \in (0,1)$, and let $x_{H,1} < x_{H,2} < \cdots$ be the positive real zeros of the Bessel function J_{1-H}. For $k \in \mathbb{N}$, define

$$(\tau_{H,k})^2 = \frac{2c_H^2}{x_{H,k}^{2H} J^2_{-H}(x_{H,k})}, \qquad [3.39]$$

where

$$c_H^2 = \frac{\Gamma(1+2H)\sin \pi H}{\pi}. \qquad [3.40]$$

The following main proposition is due to [DZH 04], theorem 4.3.

PROPOSITION 3.3.– *Considering a sub-fractional Brownian motion* ξ^H*, for all* $s, t \in [0,1]$*, we have*

$$\mathbb{E}(\xi_s^H \xi_t^H) = 2\sum_{n=1}^{\infty} \frac{\Big(1-\cos(x_{H,n}s)\Big)\Big(1-\cos(x_{H,n}t)\Big)}{x_{H,n}^2}\tau_{H,n}^2,$$

where the series converges absolutely and uniformly in $(s,t) \in [0,1]^2$.

The following lemma will be useful to get the main theorem of this section.

LEMMA 3.10.– *Recalling the notation* $S_t^H(a,b) := a\xi_t^{1/2} + b\xi_t^H$*, we have*

$$\mathbb{E}(S_s^H(a,b)S_t^H(a,b)) = 2a^2\sum_{n=1}^{\infty} \frac{\Big(1-\cos(x_{1/2,n}s)\Big)\Big(1-\cos(x_{1/2,n}t)\Big)}{x_{1/2,n}^2}\tau_{1/2,n}^2$$
$$+ 2b^2\sum_{n=1}^{\infty} \frac{\Big(1-\cos(x_{H,n}s)\Big)\Big(1-\cos(x_{H,n}t)\Big)}{x_{H,n}^2}\tau_{H,n}^2,$$

for any $(s,t) \in [0,1]^2$*, where the series converges absolutely and uniformly in* $(s,t) \in [0,1]^2$.

PROOF.– The proof immediately follows from the expansion given in proposition 3.3 and from the independence of the Gaussian processes $\xi^{1/2}$ and ξ^H. □

Let us now consider two sequences of mutually independent, centered Gaussian random variables $Y_1, Y_2, \ldots$ and $Y_{H,1}, Y_{H,2}, \ldots$ on a common probability space, with

$\mathrm{Var}(Y_n) = \tau^2_{1/2,n}$ and $\mathrm{Var}(Y_{H,n}) = \tau^2_{H,n}$, whose values are taken from equation [3.39]. From lemma 3.10, we get the following statement.

COROLLARY 3.8.– *If we denote by X^n the partial sum process defined by*

$$X_t^n = a \sum_{k=1}^{n} \frac{1 - \cos(x_{1/2,k}t)}{x_{1/2,k}} Y_k + b \sum_{k=1}^{n} \frac{1 - \cos(x_{H,k}t)}{x_{H,k}} Y_{H,k}, \qquad [3.41]$$

then the finite-dimensional distributions (fdds) of the process X^n converge weakly to the fdds of an msfBm $S_t^H(a,b) := a\xi_t^{1/2} + b\xi_t^H$.

PROOF.– For any $n \in \mathbb{N}$, $\{X_t^n, t \in [0,1]\}$ is a centered Gaussian process, and

$$\mathrm{Cov}(X_s^n, X_t^n) = 2a^2 \sum_{k=1}^{n} \frac{\Big(1 - \cos(x_{1/2,k}s)\Big)\Big(1 - \cos(x_{1/2,k}t)\Big)}{x^2_{1/2,k}} \tau^2_{1/2,k}$$
$$+ 2b^2 \sum_{k=1}^{n} \frac{\Big(1 - \cos(x_{H,k}s)\Big)\Big(1 - \cos(x_{H,k}t)\Big)}{x^2_{H,k}} \tau^2_{H,k}.$$

Therefore, it follows from lemma 3.10 that $\lim_{n\to\infty} \mathrm{Cov}(X_s^n, X_t^n) = \mathrm{Cov}(S_s^H, S_t^H)$. Consider any integer number $p > 0$, $(t_1, \ldots, t_p) \in [0,1]$ and the sequence of centered Gaussian vectors $\{(X_{t_1}^n, \ldots, X_{t_p}^n), n \in \mathbb{N}\}$. For any $n \geq 1$, introduce the covariance matrix

$$K_n = \Big(\mathrm{Cov}(X_{t_i}^n, X_{t_j}^n)\Big)_{1 \leq i,j \leq p}.$$

Then $(X_{t_1}^n, \ldots, X_{t_p}^n) \sim \mathcal{N}(0, K_n)$, and from lemma 1.5, its characteristic function is given by

$$\varphi_n(\lambda) = \exp\Big(-\frac{1}{2}\lambda^T K_n \lambda\Big).$$

Because $\lim_{n\to\infty} \mathrm{Cov}(X_{t_i}^n, X_{t_j}^n) = \mathrm{Cov}(S_{t_i}^H, S_{t_j}^H)$, we have that $\varphi_n(\lambda)$ tends to

$$\varphi(\lambda) = \exp\Big(-\frac{1}{2}\lambda^T K \lambda\Big),$$

as $n \to \infty$, where K denotes the covariance matrix of the centered Gaussian vector $(S_{t_1}^H, \ldots, S_{t_p}^H)$, which implies the stated result. □

The main result of this section is the following theorem.

THEOREM 3.7.– *Let a and b be non-zero real numbers, and stochastic process $X^H(a,b) = (X_t^H(a,b))_{t\in[0,1]}$ defined by the equality*

$$X_t^H(a,b) = a\sum_{n=1}^{\infty}\frac{1-\cos(x_{1/2,n}t)}{x_{1/2,n}}Y_n + b\sum_{n=1}^{\infty}\frac{1-\cos(x_{H,n}t)}{x_{H,n}}Y_{H,n} \qquad [3.42]$$

is well defined, and with probability 1 both series on the right-hand side of equation [3.42] converge absolutely and uniformly in $t \in [0,1]$. The process $X^H(a,b)$ is a mixed sub-fractional Brownian motion.

PROOF.– Let $C[0,1]$ be the space of continuous functions on $[0,1]$, endowed with the supremum metric. Consider the sequence $\{X^n, n \geq 1\}$ of processes defined by equation [3.41]. It follows from theorem A.17 that the sequence $\{X^n, n \geq 1\}$ converges in $C[0,1]$ with probability 1 if and only if it has a weak limit in $C[0,1]$. In turn, by virtue of theorem A.16 and lemma 3.10, it remains to prove that the sequence $\{X^n, n \geq 1\}$ is tight in $C([0,1])$. In order to prove this, consider the terms

$$X_{1,t}^n = a\sum_{k=1}^{n}\frac{1-\cos(x_{1/2,k}t)}{x_{1/2,k}}Y_k \text{ and } X_{2,t}^n = b\sum_{k=1}^{n}\frac{1-\cos(x_{H,k}t)}{x_{H,k}}Y_{H,k},$$

and treat the second one. The tightness of the first term can be shown exactly in the same manner.

Recall that the process $X_{2,\cdot}^n$ is Gaussian. Therefore, if we denote by d_n the semi-metric defined by the equality

$$d_n^2(s,t) = \mathrm{Var}(X_{2,t}^n - X_{2,s}^n) = b^2\sum_{k=1}^{n}\frac{(\cos(x_{H,k}t) - \cos(x_{H,k}s))^2}{x_{H,k}^2}\tau_{H,k}^2,$$

then, according to theorem 1.4, (ii), we have

$$\mathbb{E}\sup_{d_n(s,t)\leq\delta}|X_{2,t}^n - X_{2,s}^n| \leq C\int_0^{\delta}\sqrt{\log N(\epsilon, d_n, [0,1])}d\epsilon, \qquad [3.43]$$

for any $\delta \in (0,1)$, where C is a universal constant, and $N(\epsilon, d_n, [0,1])$ is the minimal number of closed balls of radius ϵ needed to cover $[0,1]$. We get from inequalities [3.11] that

$$d_n(s,t) \leq d_0(s,t) := M|t-s|^{1/2\wedge H},$$

where $M = \sqrt{a^2 + b^2\nu}$ with $\nu = \begin{cases} 1 \text{ if } H > 1/2 \\ 2 - 2^{2H-1} \text{ if } H \leq 1/2 \end{cases}$. Consequently,

$$N(\epsilon, d_n, [0,1]) \leq N(\epsilon, d_0, [0,1]).$$

Taking into account remark 1.12 and combining it with [3.43], we deduce that for sufficiently small $\delta > 0$

$$\mathbb{E} \sup_{d_n(s,t)\leq\delta} |X^n_{2,t} - X^n_{2,s}| \leq C \int_0^\delta \sqrt{\log N(\epsilon, d_0, [0,1])} d\epsilon$$
$$\leq C\,(2 \vee H^{-1}) \int_0^\delta \sqrt{\log\left(\frac{M}{\epsilon}\right)} d\epsilon,$$

because the number of balls of d_0-radius ϵ that are needed to cover $[0,1]$ is bounded by $\left(\frac{M}{2\epsilon}\right)^{2\vee H^{-1}} + 1$.

The integral on the right-hand side converges to 0 as $\delta \searrow 0$, so it follows that the processes $X^n_{2,\cdot}$ are uniformly equicontinuous in probability; hence $X^n_{2,\cdot}$ is tight in $C([0,1])$ (see theorem A.18 in the Appendix). Thus, we have that partial sums of X^n converge weakly in $C[0,1]$. □

3.5.2. *Rate of convergence*

With the notations already introduced in theorem 3.7 and in lemma 3.10, in this section, we investigate the rate at which the partial sum process X^n approaches the process X^H, defined in equation [3.42]. First, we consider the covariance functions, for which we have the following result.

PROPOSITION 3.4.– *For any $H \in (0,1)$, we have that*

$$\limsup_{n\to\infty} n^{2H\wedge 1} \sup_{s,t\in[0,1]} \left|\mathbb{E}(X^H_s X^H_t) - \mathbb{E}(X^n_s X^n_t)\right| \leq 8a^2 \frac{c^2_{1/2}}{\pi} + 4b^2 \frac{c^2_H}{H\pi^{2H}},$$

where c^2_H is defined by equation [3.40].

PROOF.– On the one hand, we get from theorem 3.7 and equality [3.41] that

$$\sup_{s,t\in[0,1]} \left|\mathbb{E}(X^H_s X^H_t) - \mathbb{E}(X^n_s X^n_t)\right|$$
$$= \sup_{s,t\in[0,1]} \left(2a^2 \sum_{k>n} \frac{\left(1-\cos(x_{1/2,k}s)\right)\left(1-\cos(x_{1/2,k}t)\right)}{x^2_{1/2,k}} \tau^2_{1/2,k} \right.$$
$$\left. + 2b^2 \sum_{k>n} \frac{\left(1-\cos(x_{H,k}s)\right)\left(1-\cos(x_{H,k}t)\right)}{x^2_{H,k}} \tau^2_{H,k} \right)$$
$$\leq 8\left(a^2 \sum_{k>n} \frac{\tau^2_{1/2,k}}{x^2_{1/2,k}} + b^2 \sum_{k>n} \frac{\tau^2_{H,k}}{x^2_{H,k}} \right)$$

where in the last inequality we used the fact that

$$\left(1 - \cos(x_{1/2,k}s)\right)\left(1 - \cos(x_{1/2,k}t)\right) \le 4.$$

On the other hand, applying assertion (ii) of lemma 3.9 with $\nu = 1 - H$, we get that for any $k \ge 1$

$$J_{-H}(x_{H,k}) = J_{\nu-1}(x_{H,k}) = -J_{\nu+1}(x_{H,k}).$$

Therefore, applying assertion (i) of lemma 3.9, we obtain that

$$J_{-H}^2(x_{H,k}) = J_{\nu+1}^2(x_{H,k}) \sim \frac{2}{k\pi^2} \text{ and } x_{H,k} \sim k\pi \text{ as } k \to \infty.$$

Consequently,

$$(\tau_{H,k})^2 = \frac{2c_H^2}{x_{H,k}^{2H} J_{-H}^2(x_{H,k})} \sim \frac{c_H^2}{\pi^{2H-2}} k^{1-2H} \text{ as } k \to \infty,$$

which yields

$$\sum_{k>n} \frac{\tau_{H,k}^2}{x_{H,k}^2} \sim \frac{c_H^2}{\pi^{2H}} \sum_{k>n} \frac{1}{k^{1+2H}} \text{ and } \sum_{k>n} \frac{\tau_{1/2,k}^2}{x_{1/2,k}^2} \sim \frac{c_{1/2}^2}{\pi} \sum_{k>n} \frac{1}{k^2} \text{ as } n \to \infty. \quad [3.44]$$

Furthermore, for any $H \in (0,1)$

$$\sum_{k>n} \frac{1}{k^{1+2H}} \sim \int_n^\infty \frac{1}{x^{1+2H}} dx = \frac{1}{2Hn^{2H}}, \quad [3.45]$$

whence

$$a^2 \sum_{k>n} \frac{\tau_{1/2,k}^2}{x_{1/2,k}^2} + b^2 \sum_{k>n} \frac{\tau_{H,k}^2}{x_{H,k}^2} \sim a^2 \frac{c_{1/2}^2}{\pi} \frac{1}{n} + b^2 \frac{c_H^2}{2H\pi^{2H}} \frac{1}{n^{2H}}, \quad [3.46]$$

as $n \to \infty$, and the proof is complete. □

In the following theorem, we specify the rate of convergence of the sample paths of the partial sum process X^n to the sample paths of the msfBm X^H.

THEOREM 3.8.– *For all $H \in (0,1)$ we have*

$$i) \quad \limsup_{n\to\infty} n^{2H\wedge 1} \sup_{t\in[0,1]} \mathbb{E}(X_t^n - X_t^H)^2 \le 4a^2 \frac{c_{1/2}^2}{\pi} + 2b^2 \frac{c_H^2}{H\pi^{2H}},$$

$$ii) \quad \limsup_{n\to\infty} n^{(1-\epsilon)(H\wedge 1/2)}\mathbb{E}\sup_{t\in[0,1]}|X_t^n - X_t^H| < \infty, \ \forall\epsilon > 0,$$

where c_H^2 is defined by equation [3.40].

PROOF.– Lets us prove (i). Observe that

$$X_t^H - X_t^n = a\sum_{k>n}\frac{1-\cos(x_{1/2,k}t)}{x_{1/2,k}}Y_k + b\sum_{k>n}\frac{1-\cos(x_{H,k}t)}{x_{H,k}}Y_{H,k},$$

hence

$$\mathbb{E}(X_t^H - X_t^n)^2 \leq 4a^2\sum_{k>n}\frac{\tau_{1/2,k}^2}{x_{1/2,k}^2} + 4b^2\sum_{k>n}\frac{\tau_{H,k}^2}{x_{H,k}^2}.$$

Now the statement (i) follows from the relation [3.46].

In order to prove (ii), introduce the notations

$$R_t^{n,H} = \sum_{k>n}\frac{1-\cos(x_{H,k}t)}{x_{H,k}}Y_{H,k} \text{ and } R_t^n = \sum_{k>n}\frac{1-\cos(x_{1/2,k}t)}{x_{1/2,k}}Y_k.$$

Then, we can write that

$$X_t^H - X_t^n = aR_t^n + bR_t^{n,H}.$$

Consider the semi-metric d_n defined by the equality

$$d_n^2(s,t) = \mathrm{Var}(R_t^{n,H} - R_s^{n,H}).$$

It is easy to observe that for all $p, q > 0$

$$(d_n(s,t))^{2(p+q)} = \left|\sum_{k>n}\frac{(\cos(x_{H,k}t)-\cos(x_{H,k}s))^2}{x_{H,k}^2}\tau_{H,k}^2\right|^{p+q}$$

$$= \left|\sum_{k>n}\frac{(\cos(x_{H,k}t)-\cos(x_{H,k}s))^2}{x_{H,k}^2}\tau_{H,k}^2\right|^{p}\left|\sum_{k>n}\frac{(\cos(x_{H,k}t)-\cos(x_{H,k}s))^2}{x_{H,k}^2}\tau_{H,k}^2\right|^{q}.$$

On the one hand, we have

$$(\cos(x_{H,k}t) - \cos(x_{H,k}s))^2 \leq 4.$$

On the other hand, it follows from equalities [3.42] and $\mathrm{Var}(Y_{H,k}) = \tau_{H,k}^2$ that

$$\sum_{k>n} \frac{(\cos(x_{H,k}t) - \cos(x_{H,k}s))^2}{x_{H,k}^2}\tau_{H,k}^2 \leq \sum_{k=0}^{+\infty} \frac{(\cos(x_{H,k}t) - \cos(x_{H,k}s))^2}{x_{H,k}^2}\tau_{H,k}^2$$

$$\leq \frac{1}{b^2}\mathbb{E}(X_t^H - X_s^H)^2.$$

Therefore,

$$d_n(s,t)^{2(p+q)} \leq \frac{4^p}{b^{2q}}\left|\sum_{k>n} \frac{\tau_{H,k}^2}{x_{H,k}^2}\right|^p \left|\mathbb{E}(X_t^H - X_s^H)^2\right|^q.$$

Taking into account inequalities [3.11], we deduce that

$$\mathbb{E}(X_t^H - X_s^H)^2 \leq (a^2 + b^2\nu)|t-s|^{1\wedge 2H};\ \nu = \begin{cases} 1 \text{ if } H > 1/2, \\ 2 - 2^{2H-1} \text{ if } H \leq 1/2. \end{cases}$$

Thus, denoting $a(n) = \sum_{k>n} \frac{\tau_{H,k}^2}{x_{H,k}^2}$, we get that

$$d_n(s,t) \leq \widehat{d}_n(s,t) := \widehat{a}(n)|t-s|^\beta \qquad [3.47]$$

where

$$\widehat{a}(n) = \frac{4^p}{b^{2q}}(a^2 + b^2\nu)^{\frac{q}{2(p+q)}}(a(n))^{\frac{p}{2(p+q)}} \text{ and } \beta = (1/2 \wedge H)\frac{q}{p+q}.$$

Therefore,

$$N(\epsilon, d_n, [0,1]) \leq N\left(\epsilon, \widetilde{d}_n, [0,1]\right) \leq \left(\frac{\widehat{a}(n)}{2\epsilon}\right)^{1/\beta} + 1. \qquad [3.48]$$

It follows from theorem 1.4 that

$$\mathbb{E}(\sup_{t\in[0,1]} |R_t^{n,H}|) \leq C\int_0^{D/2} \sqrt{\log N(\epsilon, d_n, [0,1])}d\epsilon,$$

where C is some universal positive constant and D is the diameter of $[0,1]$ with respect to d_n. Considering relation [3.47], we clearly get that $D \leq \widehat{a}(n)$. Therefore,

$$\mathbb{E}(\sup_{t\in[0,1]} |R_t^{n,H}|) \leq C\int_0^{D/2} \sqrt{\log N(\epsilon, d_n, [0,1])}d\epsilon$$
$$\leq C\int_0^{\widehat{a}(n)} \left(\log N(\epsilon, \widehat{d}_n, [0,1])\right)^{1/2} d\epsilon$$
$$\leq C\int_0^{\widehat{a}(n)} \left(\log\left(\left(\frac{\widehat{a}(n)}{2\epsilon}\right)^{\frac{1}{\beta}} + 1\right)\right)^{1/2} d\epsilon,$$

where for obtaining the last inequality, we used relations [3.48].

Applying the change of variables $z = \log\left(\left(\frac{\widehat{a}(n)}{2\epsilon}\right)^{\frac{1}{\beta}} + 1\right)$, we conclude that

$$\mathbb{E}\sup_{t\in[0,1]} |R_t^{n,H}| \leq C\frac{\widehat{a}(n)\beta}{2}\int_{\log 2}^{\infty} z^{1/2}e^z(e^z-1)^{-\beta-1}dz.$$

Taking equations [3.44] and [3.45] into account, we obtain that $a(n) \sim \frac{c_H{}^2}{2H\pi^{2H}}n^{-2H}$ as $n\to\infty$, and consequently,

$$\limsup_{n\to\infty} n^{\frac{p}{p+q}H}\mathbb{E}\sup_{t\in[0,t]} |R_t^{n,H}| < \infty. \qquad [3.49]$$

By applying the same technique, we get

$$\limsup_{n\to\infty} n^{\frac{p}{2(p+q)}}\mathbb{E}\sup_{t\in[0,1]} |R_t^n| < \infty. \qquad [3.50]$$

Relations [3.49] and [3.50] yield

$$\limsup_{n\to\infty} n^{\frac{p}{p+q}(1/2\wedge H)}\mathbb{E}\sup_{t\in[0,1]} |X_t^n - X_t^H| < \infty,$$

for arbitrary positive numbers p and q, that completes the proof. □

REMARK 3.8.– Theorem 3.8 states that the values of the partial sum process X^n tend to the values of a mixed sub-fractional Brownian motion at rate $n^{H\wedge 1/2}$ in $L^2(\Omega,\mathcal{F},\mathbb{P})$, and the trajectories of the partial sum process X^n tend to the trajectories of a mixed sub-fractional Brownian motion at least at rate $n^{(H\wedge 1/2)(1-\epsilon)}$ for any $\epsilon > 0$. The fact that both of the rates increase with H can be explained by the nature of the trajectories of the msfBm. When H becomes smaller, the trajectories of the msfBm fluctuate more rapidly. Hence, we should indeed expect that for smaller H, we need more terms in the series [3.41] to achieve a given level of accuracy of the approximation.

3.5.3. *Computer simulation of msfBm sample paths*

The simulation of the trajectories of a mixed sub-fractional Brownian motion is the main application of the previous expansion. The reason why this method is particularly interesting for simulation purposes is that efficient algorithms exist to compute the zeros of Bessel functions. Moreover, these zeros only have to be computed once, regardless of the number of traces that need to be simulated. In order to perform the simulations, first, we truncated the expansion obtained in theorem 3.7 at the level $n = 2000$. Then, using Matlab® software packages, we inserted numerical values of the positive real zeros of the Bessel function J_{1-H}, for Hurst parameters $H = 0.25, H = 0.5$ and $H = 0.75$. Finally, we simulated the msfBm trajectories and obtained Figures 3.1 and 3.2. The simulation results with different valuations of H; a and b illustrate the main property of msfBm: a large value of H corresponds to smoother trajectories. In other words, for smaller values of H, the trajectories of an msfBm fluctuate more wildly. In particular, in Figure 3.2, we present three trajectories of the sub-fractional Brownian motion ξ^H, corresponding to the msfBm $S^H(a,b) = a\xi^{1/2} + b\xi^H$, with $a = 0$ and $b = 1$.

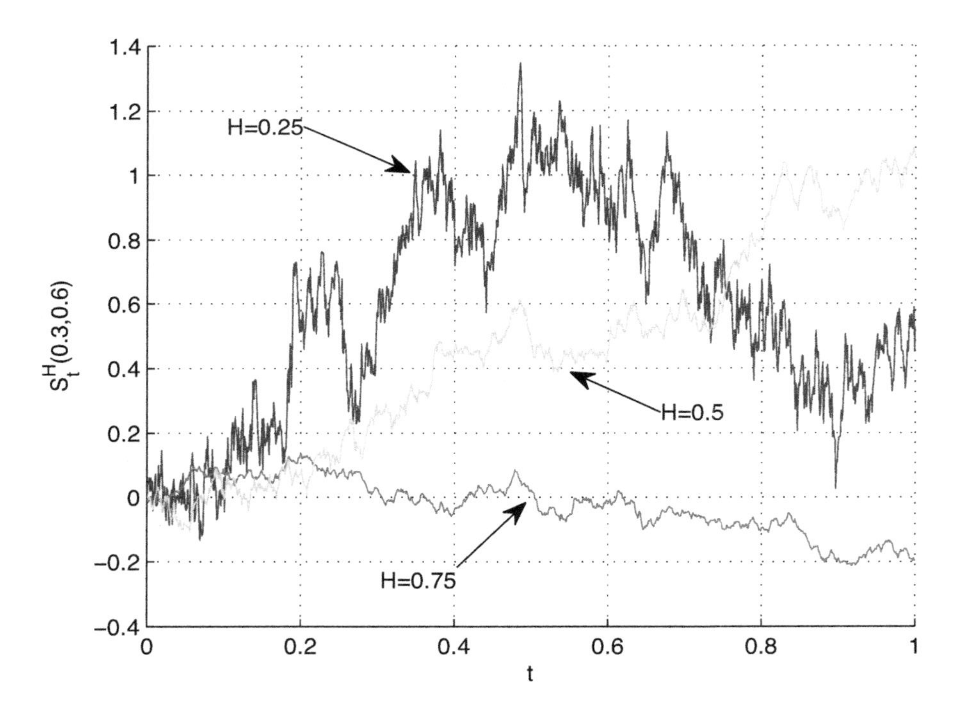

Figure 3.1. *Generated smfBm sample paths for different values of $H(a = 0.3, b = 0.6)$. For a color version of this figure, see www.iste.co.uk/mishura/stochastic.zip*

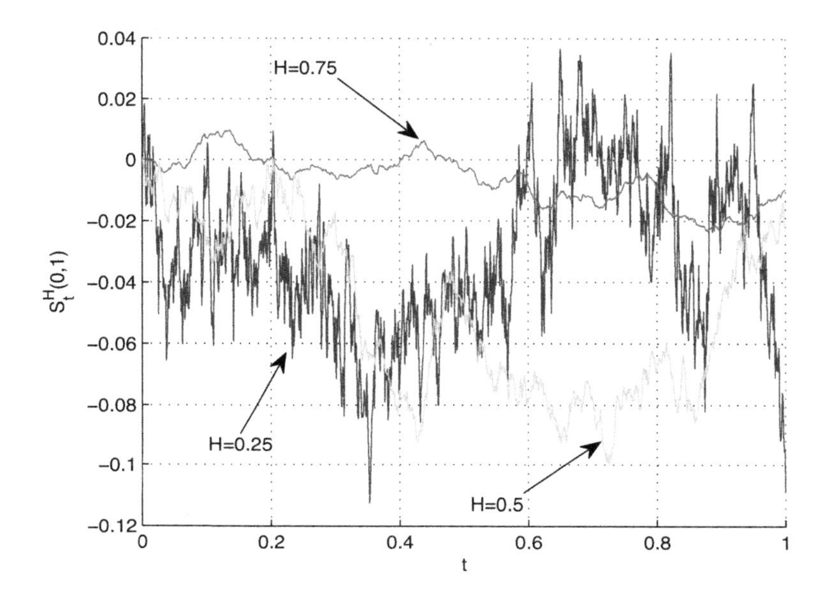

Figure 3.2. *Generated msfBm sample paths for different values of $H(a = 0, b = 1)$. For a color version of this figure, see www.iste.co.uk/mishura/stochastic.zip*

3.6. Study of the semi-martingale property of the mixed processes

3.6.1. *General properties of semi-martingales*

Consider a probability space $(\Omega, \mathcal{F}, \mathbb{P})$, and let $\mathbb{F} = \{\mathcal{F}_t, t \geq 0\}$ be a filtration defined on it (see definition A.30 in the Appendix). Also, let $X = \{X_t, t \geq 0\}$ be a real-valued stochastic process.

DEFINITION 3.2.– *A stochastic process X is adapted to $\mathbb{F}$ (or $\mathbb{F}$-adapted), if the random variable X_t is $\mathcal{F}_t$-measurable for any $t \geq 0$.*

EXAMPLE 3.1.– *Any stochastic process X is adapted to its natural filtration $\mathcal{F}^X$ (see example A.7 in the Appendix).*

DEFINITION 3.3.– *A stochastic process is a martingale with respect to $\mathbb{F}$ (or $\mathbb{F}$-martingale), if the following assertions are fulfilled:*

i) $\mathbb{E}(|X_t|) < \infty$ *for any* $t \geq 0$.

ii) $\mathbb{E}(X_t|\mathcal{F}_s) = X_s$ *for any* $0 \leq s \leq t$.

REMARK 3.9.– From the second assertion of definition 3.3, we see that if X is a martingale with respect to $\mathbb{F}$, then X is necessarily $\mathbb{F}$-adapted.

DEFINITION 3.4.– *Consider a measurable space* $(\Omega, \mathcal{F})$ *equipped with a filtration* $\mathbb{F} = \{\mathcal{F}_t, t \geq 0\}$. *A stopping time of the filtration* $\mathbb{F}$ *is a random variable* $\tau \geq 0$ *such that the event* $\{\tau \leq t\} \in \mathcal{F}_t$ *for any* $t \geq 0$.

DEFINITION 3.5.– *A process* X *is called a local martingale with respect to* $\mathbb{F}$, *if there exists a non-decreasing sequence* $\{\tau_n, n \geq 1\}$ *of stopping times on* $\mathbb{F}$ *such that*

i) $\lim_{n \to \infty} \tau_n = \infty$ *a.s.*

ii) A stochastic process $\{X^{\tau_n}, t \geq 0\} \equiv \{X_{t \wedge \tau_n}, t \geq 0\}$ *is an* $\mathbb{F}$*-martingale for any* $n \geq 1$.

LEMMA 3.11.– *Consider a continuous local martingale* M. *Then,* M *is of finite quadratic variation, and moreover,* $[M]_t = 0, \forall t \geq 0$ *if and only if* $M_t = M_0$ *a.s.*, $\forall t \geq 0$.

For the proof of lemma 3.11 see, for example, proposition 1.12 and theorem 1.18 [REV 91].

DEFINITION 3.6.– *Suppose that the filtration* $\mathbb{F}$ *satisfies the standard assumptions. The process* X *is called a semi-martingale with respect to* $\mathbb{F}$, *if the following assertions are fulfilled:*

1) X *is adapted with respect to* $\mathbb{F}$.

2) X *admits the decomposition*

$$X_t = M_t + A_t, \quad t \geq 0,$$

where M *is a right-continuous local martingale with respect to* $\mathbb{F}$ *and* A *is a right-continuous and* $\mathbb{F}$*-adapted process of finite variation (see definition A.35 in the Appendix).*

EXAMPLE 3.2.– *Any* $\mathbb{F}$*-local martingale is a semi-martingale with respect to* $\mathbb{F}$, *and any* $\mathbb{F}$*-adapted process of finite variation is a semi-martingale with respect to* $\mathbb{F}$. Consider a quadratic variation of a semi-martingale.

LEMMA 3.12.– *A semi-martingale* $X = M + A$ *with continuous components* M *and* A *has a finite quadratic variation and, moreover, for any* $t \geq 0$, *it holds that* $[X]_t = [M]_t$.

PROOF.– For any $t > 0$ and any partition $\pi_m = \{t_0, t_1, \ldots, t_m\}$ of $[0, t]$, with $0 = t_0 < t_1 < \ldots < t_m = t$, we have

$$V_t^{(2)}(\pi_m) = \sum_{k=1}^{m} |X_{t_k} - X_{t_{k-1}}|^2$$

$$= \sum_{k=1}^{m} |M_{t_k} - M_{t_{k-1}}|^2 + \sum_{k=1}^{m} |A_{t_k} - A_{t_{k-1}}|^2$$

$$+ 2\sum_{k=1}^{m} (M_{t_k} - M_{t_{k-1}})(A_{t_k} - A_{t_{k-1}}).$$

Because

$$\Big|\sum_{k=1}^{m} (M_{t_k} - M_{t_{k-1}})(A_{t_k} - A_{t_{k-1}})\Big| \leq \Big(\sup_{1\leq k\leq m} |M_{t_k} - M_{t_{k-1}}|\Big) V_t(A),$$

where $V_t(A)$ is the variation of A on $[0, t]$, which is finite (see definition A.1 in the Appendix), and because

$$\lim_{|\pi_m|\to 0} \Big(\sup_{1\leq k\leq m} |M_{t_k} - M_{t_{k-1}}|\Big) = 0,$$

(which is due to the continuity of M), we get

$$\lim_{|\pi_m|\to 0} \sum_{k=1}^{m} (M_{t_k} - M_{t_{k-1}})(A_{t_k} - A_{t_{k-1}}) = 0,$$

and consequently,

$$[X]_t = [M]_t + [A]_t = [M]_t, t \geq 0$$

where in the last equality, we used lemma A.13 from the Appendix. □

The following theorem is known as the Bichteler–Dellacherie theorem. Its proof can be found, for example, in [DEL 80], section 8.4.

In order to formulate the theorem, let us introduce the following class of random variables. Namely, we put

$$\theta(\mathbb{F}) = \Bigg\{\theta = \sum_{j=0}^{n-1} f_j \mathbf{1}_{(t_j, t_{j+1}]},\ n \in \mathbb{N},\ 0 = t_0 \leq \ldots \leq t_n = T,$$

$$f_j \text{ is } \mathcal{F}_{t_j}\text{-measurable and } |f_j| \leq 1 \text{ a.s. for any } 0 \leq j \leq n-1\Bigg\}.$$

For any $\theta \in \theta(\mathbb{F})$, we put

$$I_X(\theta) = \sum_{j=0}^{n-1} f_j \left(X_{t_{j+1}} - X_{t_j}\right). \qquad [3.51]$$

THEOREM 3.9.– *Assume that for some $T > 0$, a filtration $\mathbb{F} = \{\mathcal{F}_t, 0 \le t \le T\}$ satisfies the standard assumptions. An a.s. right-continuous, $\mathbb{F}$-adapted stochastic process $\{X_t, 0 \le t \le T\}$ is a $\mathbb{F}$-semi-martingale if and only if the set of integral sums*

$$I_X(\theta(\mathbb{F})) := \{I_X(\theta),\ \theta \in \theta(\mathbb{F})\}$$

is bounded in the space $L^0(\Omega, \mathcal{F}, \mathbb{P})$ of finite random variables.

If we do not require the process to be a.s. right continuous and the filtration to satisfy the standard assumptions, we present a weaker form of the semi-martingale than the one defined above.

DEFINITION 3.7.– *A stochastic process $\{X_t, 0 \le t \le T\}$ is a weak semi-martingale with respect to a filtration $\mathbb{F} = \{\mathcal{F}_t, 0 \le t \le T\}$ if X is $\mathbb{F}$-adapted and $I_X(\theta(\mathbb{F}))$ is bounded in $L^0(\Omega, \mathcal{F}, \mathbb{P})$.*

REMARKS 3.9.–

1) If the filtration $\mathbb{F} = \{\mathcal{F}_t, 0 \le t \le T\}$ satisfies the standard assumptions and the process X is a.s. right-continuous, then X is a $\mathbb{F}$-semi-martingale if and only if X is a weak semi-martingale with respect to $\mathbb{F}$.

2) Let $X = \{X_t, 0 \le t \le T\}$ be a stochastic process. If $\mathbb{F}^1 = \{\mathcal{F}^1_t, 0 \le t \le T\}$ and $\mathbb{F}^2 = \{\mathcal{F}^2_t, 0 \le t \le T\}$ are two filtrations such that $\mathcal{F}^1_t \subset \mathcal{F}^2_t$ for any $t \in [0, T]$, then $\theta(\mathbb{F}^1) \subset \theta(\mathbb{F}^2)$. Therefore, L^0-boundedness of $I_X\left(\theta(\mathbb{F}^2)\right)$ implies the L^0-boundedness of $I_X\left(\theta(\mathbb{F}^1)\right)$. This shows that if X is not a weak semi-martingale with respect to its natural filtration, then it is not a weak semi-martingale with respect to any other filtration to which it is adapted. Therefore, it is natural that in the following sections of this book, we will adopt the following definition.

DEFINITION 3.8.– *Let $\{X_t, 0 \le t \le T\}$ be a stochastic process. We call X a weak semi-martingale if it is a weak semi-martingale with respect to its natural filtration $\mathbb{F}^X = \{\mathcal{F}^X_t, 0 \le t \le T\}$. We call X a semi-martingale if it is a semi-martingale with respect to $\overline{\mathcal{F}}^X$, the smallest filtration that contains $\mathcal{F}^X$ and satisfies the standard assumptions.* The following theorem is due to [STR 84].

THEOREM 3.10.– *Consider a Gaussian process $X = \{X_t, 0 \le t \le T\}$. Let $\mathbb{F}^X$ be its natural filtration and $I_X\left(\theta(\mathbb{F}^X)\right)$ be the collection of random variables defined by* *equation [3.51]. If* $I_X\left(\theta(\mathbb{F}^X)\right)$ *is bounded in* $L^0(\Omega,\mathcal{F},\mathbb{P})$*, then it is also bounded in* $L^2(\Omega,\mathcal{F},\mathbb{P})$.

3.6.2. *Semi- and non-semi-martingale property for the mixed sub-fractional Brownian motion*

In this section, we discuss for which values of the Hurst parameter H, $S^H(N,a)$ is a semi-martingale. The study of this question in the case of $M^H(N,a)$ is similar.

THEOREM 3.11.– *For any* $t>0$ *denote by* $[S^H(N,a)]_t$ *(respectively* $V(S^H)_t$*) the quadratic variation (respectively the variation) of the process* $S^H(N,a)$ *on the interval* $[0,t]$*, see definitions A.35 and A.36 in the Appendix. Then, the following statements hold.*

1) Let $H_1<1/2$*. Then,* $[S^H(N,a)]_t=+\infty$ *a.s.*

2) Let $H_1>1/2$*. Then,*

i) $[S^H(N,a)]_t=0$ *a.s.*

ii) $V(S^H)_t=\infty$ *a.s.*

PROOF.– (1) For any $n\in\mathbb{N}$, $p\in\mathbb{N}$ and $t>0$, denote

$$A_{n,p}=\sum_{j=1}^{n}\left|S^H_{\frac{jt}{n}}(N,a)-S^H_{\frac{(j-1)t}{n}}(N,a)\right|^p.$$

Assume that for some $t>0$ $[S^H(N,a)]_t<\infty$ a.s. Then, $A_{n,2}\to[S^H(N,a)]_t$ in probability as $n\to\infty$, so there is a subsequence $\{n_k,k\geq 1\}$ such that $A_{n_k,2}\to[S^H(N,a)]_t$ a.s. as $k\to\infty$ and therefore $\sup_k A_{n_k,2}<\infty$ a.s. Let

$$v(x)=\left(\sup_k\sum_{j=1}^{n_k}\left[x\left(\frac{jt}{n_k}\right)-x\left(\frac{(j-1)t}{n_k}\right)\right]^2\right)^{1/2},\ x\in C([0,t]).$$

Then, v is a pseudo-semi-norm on $C([0,t])$ such that $v(S^H(N,a))<\infty$ *a.s.*, see example A.5 in the Appendix. Therefore, by virtue of theorem A.15 of the Appendix, there exists $\epsilon>0$ such that for any $\alpha<\epsilon$

$$\mathbb{E}\exp(\alpha v^2(S^H(N,a))<\infty.$$

In turn, it means that

$$\mathbb{E}(v^4(S^H(N,a)))\leq C\mathbb{E}\exp\left(\frac{\epsilon}{2}v^2(S^H(N,a)\right)<\infty,$$

where C denotes a positive constant, depending only on ϵ.

Now, let us choose $p > 2$ such that $H_1 < \frac{1}{p} < \frac{1}{2}$. Then,

$$\mathbb{E}(A_{n_k,p}) \leq \mathbb{E}\left(\max_{j \leq n_k} \left| S^H_{j/n_k}(N,a) - S^H_{(j-1)/n_k}(N,a) \right|^{p-2} v(S^H(N,a))^2 \right)$$

$$\leq \left[\mathbb{E}\left(\max_{j \leq n_k} \left| S^H_{j/n_k}(N,a) - S^H_{(j-1)/n_k}(N,a) \right|^{2(p-2)} \right) \right]^{1/2} \left[\mathbb{E}\left(v\left(S^H(N,a) \right)^4 \right) \right]^{1/2}.$$

Because $p > 2$, the last expression tends to 0 as $k \to \infty$; indeed, the function $x \longmapsto \sup_{s \leq t} |x(s)|$ is a norm on $C([0,t])$; see example A.3.2 in the Appendix. Hence, by virtue of theorem A.15 of the Appendix, we get

$$\mathbb{E}\left(\sup_{s \leq t} |S^H_s(N,a)|^{2(p-2)} \right) < \infty,$$

and in turn, using continuity of $S^H(N,a)$ and the Lebesgue dominated convergence theorem, we get

$$\mathbb{E}\left(\max_{j \leq n_k} \left| S^H_{j/n_k}(N,a) - S^H_{(j-1)/n_k}(N,a) \right|^{2(p-2)} \right) \to 0 \text{ as } k \to \infty.$$

Because the increments of the process $S^H(N,a)$ are Gaussian, we have that

$$\begin{aligned} \mathbb{E}\Big(A_{n,p}\Big) &= \sum_{j=1}^{n} \mathbb{E}\left(\left| S^H_{\frac{jt}{n}}(N,a) - S^H_{\frac{(j-1)t}{n}}(N,a) \right|^p \right) \\ &= C \sum_{j=1}^{n} \left(\mathbb{E}\left(S^H_{\frac{jt}{n}}(N,a) - S^H_{\frac{(j-1)t}{n}}(N,a) \right)^2 \right)^{\frac{p}{2}}, \end{aligned}$$

where C denotes a positive constant. Then, by virtue of remark 3.3, we get

$$\mathbb{E}\Big(A_{n,p}\Big) \geq C \sum_{j=1}^{n} \left(\left(\frac{t}{n} \right)^{2H_1} \right)^{\frac{p}{2}},$$

and therefore,

$$a n^{1-pH_1} \leq \mathbb{E}\Big(A_{n,p}\Big),$$

where a is a positive constant that depends on a_1, H_1, p and t. Because $1 - pH_1 > 0$, we deduce that $\mathbb{E}(A_{n,p})$ tends to ∞ as $n \to \infty$. It contradicts to the previously received conclusion that $\mathbb{E}(A_{n_k,p}) \to 0$ as $k \to \infty$. Thus, the assumption that $[S^H(N,a)]_t < \infty$ a.s., is false.

Because $x \longmapsto [x]_t$ is a pseudo-semi-norm on $C([0,t])$ (see example A.6 in the Appendix), then it follows from Fernique's zero-one law (theorem A.14 in the Appendix) that $[S^H(N,a)]_t = \infty$ a.s.

(2) In order to check (i), let us consider a sequence

$$\pi_n = \Big\{0 = t_0 < t_1 < \cdots < t_n = t,\Big\},\ n \in \mathbb{N}$$

of finite partitions of $[0,t]$ such that its mesh $|\pi_n| = \max_{1\le i\le n} |t_i - t_{i-1}| \to 0$ as $n \to \infty$. From remark 3.3, we have that for any $v, s \in [0,t]$ such that $s \le v$,

$$C_1(v-s)^{2H_1} \le \mathbb{E}\left(S_v^H(N,a) - S_s^H(N,a)\right)^2 \le C_2(v-s)^{2H_1}, \qquad [3.52]$$

where C_1 and C_2 are two positive constants. Now, for any integer $n \ge 1$, denote

$$\Delta_t^{\pi_n} = \sum_{j=1}^{n} (S_{t_j}^H(N,a) - S_{t_{j-1}}^H(N,a))^2.$$

It follows from inequalities [3.52] that

$$\mathbb{E}(\Delta_t^{\pi_n}) \le C_2 \sum_{j=1}^{n} (t_j - t_{j-1})^{2H_1} \le C_2|\pi_n|^{2H_1-1} \sum_{j=1}^{n} (t_j - t_{j-1}) = C_2|\pi_n|^{2H_1-1} t.$$

Because $\lim_{n\to\infty} C_2|\pi_n|^{2H_1-1}t = 0$, we conclude that $\lim_{n\to\infty} \mathbb{E}(\Delta_t^{\pi_n}) = 0$. Hence, the sequence $(\Delta_t^{\pi_n})$ vanishes in probability, which yields that $[S^H(N,a)]_t = 0$ a.s.

In order to get (ii), it is sufficient to apply the same procedure as in the proof of item (1) and to use Fernique's theorems A.14 and A.15 in the Appendix, together with equality [3.10]. □

From theorem 3.11, we get the following statement.

COROLLARY 3.9.– *The msfBm $S^H(N,a)$ is not a semi-martingale in both of the following cases:*

i) If $1 \le i \le N$ exists such that $H_i < 1/2$.

ii) If, for any $1 \le i \le N$, we have that $H_i > 1/2$.

PROOF.– In case (i), we argue as follows: from the first assertion of theorem 3.11 we get that the quadratic variation of S^H is infinite. Therefore, by virtue of lemma 3.12 and from its continuity, S^H could not be a semi-martingale.

Therefore, let us check the second case where $H_i > 1/2$, for any $i \in \{1, \dots, N\}$. The process $S^H(N, a)$ is continuous. Suppose that it is a semi-martingale. Hence, $S^H(N, a)$ can be written in the form

$$S_t^H(N, a) = M_t + A_t,$$

where $M_0 = A_0 = 0$, M is an a.s. continuous local martingale with respect to $\overline{\mathcal{F}}^{S^H(N,a)}$ and A is an a.s. continuous $\overline{\mathcal{F}}^{S^H}$-adapted process of finite variation. It follows from lemma 3.12 that, for any $t \in [0, T]$,

$$0 = [S^H(N, a)]_t = [M]_t.$$

Now, from lemma 3.11, M is itself a zero process, and hence, $S^H(N, a) = A$ has finite variation. This contradicts assertion $(2), (ii)$ of theorem 3.11. □

Denoting $\Delta_j^n X = X_{\frac{jT}{n}} - X_{\frac{(j-1)T}{n}}$, for any $j \in \{1, \dots, n\}$ and any stochastic process $X = \{X_t, t \in [0, T]\}$, we will now prove the following important theorem.

THEOREM 3.12.– *Consider a centered Gaussian process $X = \{X_t, t \in [0, T]\}$, whose covariance function K_X is positive definite. Consider also $W = \{W_t, t \in [0, T]\}$ a standard Wiener process independent of X. If*

$$\sup_{n \in \mathbb{N}} n^2 \sum_{j,k=1}^{n} \left(\mathrm{Cov}(\Delta_j^n X, \Delta_k^n X)\right)^2 < \infty,$$

then the process

$$Y_t := X_t + W_t, \ t \in [0, T]$$

is, in its own filtration, a semi-martingale, which is equivalent to a Brownian motion.

PROOF.– Without loss of generality, suppose that $T = 1$. Here, we use the notion of entropy, see definition 1.15 in Chapter 1, and carry out a computation similar to that carried out in the proof of proposition 1.6 in Chapter 1. For all $n \in \mathbb{N}$, we define $Z_n : C([0, 1]) \to \mathbb{R}^n$ by

$$Z_n(\omega) = \left(\omega\left(\frac{1}{n}\right) - \omega(0), \omega\left(\frac{2}{n}\right) - \omega\left(\frac{1}{n}\right), \dots, \omega(1) - \omega\left(\frac{n-1}{n}\right)\right),$$

and $\mathcal{F}_n = \sigma(Z_n) = \sigma\left(\left\{\omega \in C([0, 1]) | Z_n(\omega) \in B\right\}, B \in \mathcal{B}(\mathbb{R}^n)\right)$. Note that $\vee_{n=1}^{\infty} \mathcal{F}_n$ is equal to the σ-field $\mathcal{C}^{[0,1]}$ generated by the cylinder subsets of $C([0, 1])$

(see definition 1.12 in Chapter 1). We denote by $\mathbb{P}_Y$ the measure induced by Y on $C([0,1])$ and by $\mathbb{P}_W$ Wiener measure, the measure induced by the Wiener process W on $C([0,1])$. For any $n \in \mathbb{N}$, denote by $\mathbb{P}_Y^n$ and $\mathbb{P}_W^n$ the restrictions of $\mathbb{P}_Y$ and $\mathbb{P}_W$ to $\mathcal{F}_n$, respectively.

For all $n \in \mathbb{N}$, Z_n is a centered Gaussian vector under both measures $\mathbb{P}_Y^n$ and $\mathbb{P}_W^n$. The covariance matrices of Z_n under $\mathbb{P}_Y^n$ and $\mathbb{P}_W^n$ equal, respectively,

$$K_{Z_n,\mathbb{P}_Y^n} = \mathbb{E}_{\mathbb{P}_Y^n}(Z_n Z_n^T) = \frac{1}{n}\mathbb{I}_n + \Sigma_n$$

and

$$K_{Z_n,\mathbb{P}_W^n} = \mathbb{E}_{\mathbb{P}_W^n}(Z_n Z_n^T) = \frac{1}{n}\mathbb{I}_n.$$

Here, $\mathbb{I}_n$ stands for the unit matrix and Σ_n is the covariance matrix of the increments of the vector

$$\left(X_{\frac{1}{n}}, X_{\frac{2}{n}} - X_{\frac{1}{n}}, \ldots, X_1 - X_{\frac{n-1}{n}}\right).$$

Because Σ_n is symmetric, by virtue of theorem A.7 in the Appendix, there exists an orthogonal $n \times n$ matrix U_n such that $U_n \Sigma_n U_n^T$ is a diagonal matrix of the form

$$D_n = \begin{pmatrix} \lambda_1^n & 0 & \ldots & \ldots & 0 \\ 0 & \lambda_1^n & 0 & \ldots & 0 \\ & & & & \\ \vdots & & & \ddots & 0 \\ 0 & \ldots & \ldots & 0 & \lambda_n^n \end{pmatrix}.$$

Denoting $L_n = \sqrt{n} U_n Z_n$, we observe that L_n is a centered Gaussian vector under both measures $\mathbb{P}_Y^n$ and $\mathbb{P}_W^n$. The covariance matrices of L_n under these two measures equal, respectively,

$$K_{L_n,\mathbb{P}_Y^n} = \mathbb{E}_{\mathbb{P}_Y^n}(L_n L_n^T) = n\left(U_n\left(\frac{1}{n}I_n + \Sigma_n\right)U_n^t\right) = \mathbb{I}_n + nD_n$$

and

$$K_{L_n,\mathbb{P}_W^n} = \mathbb{E}_{\mathbb{P}_W^n}(L_n L_n^T) = n\left(U_n\left(\frac{1}{n}I_n\right)U_n^t\right) = \mathbb{I}_n.$$

It follows from lemma 1.11 in Chapter 1 that $\mathbb{P}_Y^n$ and $\mathbb{P}_W^n$ are equivalent and, if we denote φ_n the Radon–Nikodym derivative of $\mathbb{P}_Y^n$ with respect to $\mathbb{P}_W^n$, then the entropy equals

$$\begin{aligned}\mathbf{H}(\mathbb{P}_Y^n|\mathbb{P}_W^n) &= \mathbb{E}_{\mathbb{P}_Y^n}\Bigg(\log\varphi_n(\omega)\Bigg)\\ &= \frac{1}{2}\sum_{k=1}^{n}\left(\sigma_k^2 - 1 - \log\sigma_k^2\right),\end{aligned}$$

where $\sigma_k^2 = \mathrm{Var}_{\mathbb{P}_Y^n}(L_k) = 1 + n\lambda_k^n$. For all $x \geq 0$, we have that

$$x - \log(1+x) = \int_0^x \frac{u}{1+u}du \leq \int_0^x udu = \frac{x^2}{2}.$$

Therefore,

$$\begin{aligned}\mathbf{H}(\mathbb{P}_Y^n|\mathbb{P}_W^n) &= \frac{1}{2}\sum_{k=1}^{n}\left(\sigma_k^2 - 1 - \log\sigma_k^2\right) = \frac{1}{2}\sum_{k=1}^{n}(n\lambda_k^n - \log(1+n\lambda_k^n)\\ &\leq \frac{1}{4}n^2\sum_{k=1}^{n}(\lambda_k^n)^2.\end{aligned}$$

To get the stated result, it is sufficient to prove that $\sup_n \mathbf{H}(\mathbb{P}_Y^n|\mathbb{P}_W^n) < \infty$. Indeed, in such case, from lemma 1.12 of Chapter 1 $\mathbb{P}_{Y^n}$ is absolutely continuous with respect to $\mathbb{P}_W^n$. However, from corollary 1.16 in Chapter 1, two Gaussian measures on $C([0,1])$ can only be equivalent or singular. Therefore, $\mathbb{P}_{Y^n}$ and $\mathbb{P}_W^n$ must be equivalent.

However, in order to get the relation $\sup_{n\in\mathbb{N}} \mathbf{H}(\mathbb{P}_Y^n|\mathbb{P}_W^n) < \infty$, it is sufficient to prove that

$$\sup_{n\in\mathbb{N}} n^2\sum_{k=1}^{n}(\lambda_k^n)^2 < \infty,$$

where $\lambda_1^n, \ldots, \lambda_n^n$ are the eigenvalues of the covariance matrix of the increments of process X. Because orthogonal transformation leaves the Hilbert–Schmidt norm of a matrix invariant, we have

$$\sum_{k=1}^{n}(\lambda_k^n)^2 = \sum_{j,k=1}^{n}\left(\mathrm{Cov}(\Delta_j^n X, \Delta_k^n X)\right)^2.$$

Consequently, if

$$\sup_{n\in\mathbb{N}} n^2 \sum_{j,k=1}^{n} \left(\mathrm{Cov}(\Delta_j^n X, \Delta_k^n X)\right)^2 < \infty,$$

then $\mathbb{P}_{Y^n}$ and $\mathbb{P}_W^n$ are equivalent. □

COROLLARY 3.10.– *Consider a centered Gaussian process $X = \{X_t, t \in [0,T]\}$, whose covariance function $K_X(s,t) = \mathrm{Cov}(X_s, X_t)$ satisfies the following conditions:*

i) The first partial derivatives $\frac{\partial K_X(s,t)}{\partial s}$ and $\frac{\partial K_X(s,t)}{\partial t}$ are continuous on $[0,T]^2$.

ii) K_X is twice continuously differentiable on $[0,T]^2 \setminus \{t = s \in [0,T]\}$.

iii) $\frac{\partial^2 K_X}{\partial s \partial t} \in L^2([0,T]^2)$.

Consider also a standard Wiener process $W = \{W_t, t \in [0,T]\}$, independent of X. Then, the process

$$Y_t := X_t + W_t, \; t \in [0,T]$$

is a semi-martingale in its own filtration and is equivalent in a measure to a Brownian motion.

PROOF.– For technical simplicity, consider $T = 1$. First, because of the condition that $\frac{\partial^2 K_X}{\partial s \partial t}$ is continuous except diagonal and is in $L^2([0,T]^2)$ and consequently in $L^1([0,T]^2)$, and due to condition (i), we can write that for any j, k

$$\mathrm{Cov}(\Delta_j^n X, \Delta_k^n X) = \int_{j/n}^{j+1/n} \int_{k/n}^{k+1/n} \frac{\partial^2 K_X(s,t)}{\partial s \partial t} ds dt. \qquad [3.53]$$

Indeed, it is obvious for $j \neq k$. For $j = k$,

$$\begin{aligned}
\mathrm{Cov}(\Delta_j^n X, \Delta_j^n X) &= K_X(j+1/n, j+1/n) - K_X(j, j+1/n) \\
&\quad - K_X(j+1/n, j/n) + K_X(j/n, j/n) \\
&= \int_{j/n}^{j+1/n} \frac{\partial K_X(s, j+1/n)}{\partial s} ds - \int_{j/n}^{j+1/n} \frac{\partial K_X(s, j/n)}{\partial s} ds \\
&= \int_{j/n}^{j+1/n} \left(\frac{\partial K_X(s, j+1/n)}{\partial s} ds - \frac{\partial K_X(s, j/n)}{\partial s} \right) ds
\end{aligned}$$

Furthermore, due to a continuity of partial derivative in the second variable and an existence a.e. of its integrable derivative $\frac{\partial^2 K_X}{\partial s \partial t}$, we can conclude that

$$\frac{\partial K_X(s, j+1/n)}{\partial s} ds - \frac{\partial K_X(s, j/n)}{\partial s} = \int_{j/n}^{j+1/n} \frac{\partial^2 K_X(s,t)}{\partial s \partial t} dsdt,$$

whence equation [3.53] follows.

Second, under the assumption that $\frac{\partial^2 K_X}{\partial s \partial t} \in L^2([0,T]^2)$, we can apply the Cauchy–Schwartz inequality to functions 1 and $\frac{\partial^2 K_X}{\partial s \partial t}$ and write

$$\begin{aligned}
&\sup_{n\in\mathbb{N}} n^2 \sum_{j,k=1}^{n} \Big(\mathrm{Cov}(\Delta_j^n X, \Delta_k^n X)\Big)^2 \\
&= \sup_{n\in\mathbb{N}} n^2 \sum_{j,k=1}^{n} \left(\int_{j/n}^{j+1/n} \int_{k/n}^{k+1/n} \frac{\partial^2 K_X(s,t)}{\partial s \partial t} dsdt\right)^2 \\
&\leq \sup_{n\in\mathbb{N}} n^2 \cdot n^{-2} \sum_{j,k=1}^{n} \int_{j/n}^{j+1/n} \int_{k/n}^{k+1/n} \left(\frac{\partial^2 K_X(s,t)}{\partial s \partial t}\right)^2 dsdt \\
&\leq \int_{[0,T]^2} \left(\frac{\partial^2 K_X(s,t)}{\partial s \partial t}\right)^2 dsdt < \infty,
\end{aligned}$$

and consequently, from theorem 3.12, we prove the corollary. □

In the following theorem, we treat the case where $H_1 = 1/2$, and all $H_i \in (3/4, 1), 2 \leq i \leq N$.

THEOREM 3.13.– *Let $H_1 = 1/2$, and $H_i \in (3/4, 1)$ for any $i \geq 2$. Then, for any $T > 0$, the process*

$$S^H(N,a) = \{S_t^H(N,a), t \in [0,T]\}$$

is, in its own filtration, a semi-martingale, equivalent in measure to $a_1 B$, where B is a Brownian motion.

PROOF.– The process $S^H(N,a)$ can be written as $S_t^H(N,a) = a_1\Big(W_t + X_t\Big)$, where

$$X_t = \sum_{i=2}^{N} \frac{a_i}{a_1} \xi_t^{H_i}.$$

The process X_t is Gaussian, and its covariance function

$$R(s,t) = \sum_{i=2}^{N} \frac{a_i^2}{a_1} \left(t^{2H_i} + s^{2H_i} - 1/2\left(|s+t|^{2H_i} + |t-s|^{2H_i}\right)\right),$$

is positive definite, twice continuously differentiable on $[0,T]^2 \setminus \{(s,t); t = s\}$, and for any $(s,t) \in [0,T]^2 \setminus \{(s,t); t = s\}$,

$$\frac{\partial^2 R(s,t)}{\partial s \partial t} = \sum_{i=2}^{N} \frac{a_i^2}{a_1} H_i(2H_i - 1)\left(|t-s|^{2H_i-2} - |s+t|^{2H_i-2}\right).$$

Also, the first derivatives of R are continuous on $[0,T]^2$.

Therefore, because $H_i \in \big(3/4, 1\big)$, for any $i \in \{2, \ldots, N\}$, it is easy to check that $\frac{\partial^2 R}{\partial s \partial t} \in L^2([0,T]^2)$.

On the other hand, W is a Brownian motion, independent of X. Therefore, it follows from corollary 3.10 that the process $\{W_t + X_t, t \in [0,T]\}$ is equivalent in measure to a Brownian motion, and consequently, it is, in its own filtration, a semi-martingale. □

We conclude this section by studying the remaining case, namely the case when $H_1 = 1/2$, and there exists $i > 1$ for which $H_i \in \big(1/2, 3/4\big]$. For brevity denote $\|Y\|_p = \|Y\|_{L^p(\Omega,\mathcal{F},\mathbb{P})} = (\mathbb{E}|Y|^p)^{1/p}$ for any random variable Y and $p \geq 1$. Let us first recall the definition of a quasi-martingale.

DEFINITION 3.9.– *A stochastic process $\{X_t, t \geq 0\}$ is called a quasi-martingale if the random variable X_t is integrable for all $t \geq 0$, and for any $T > 0$*

$$\sup_{\pi} \sum_{j=0}^{k-1} \left\| E\left(X_{t_{j+1}} - X_{t_j} | \mathcal{F}_{t_j}^X\right) \right\|_1 < \infty,$$

where π is the set of all finite partitions $0 = t_0 < t_1 < \ldots < t_n = T$ of $[0,T]$.

In the following key lemma, we specify the relation between quasi-martingale and semi-martingale in the case of our process $S^H(N,a)$.

LEMMA 3.13.– *If $S^H(N,a)$ is not a quasi-martingale, then it is not a semi-martingale, with respect to its own filtration.*

PROOF.– Let us assume that S^H is a semi-martingale. Under this assumption, the set $I_{S^H}\left(\beta\left(\mathcal{F}^{S^H}\right)\right)$, defined in equation [3.51], is bounded in $L^0(\Omega, \mathcal{F}, \mathbb{P})$. Then, from theorem 3.10, it is also bounded in $L^2(\Omega, \mathcal{F}, \mathbb{P})$ and therefore in $L^1(\Omega, \mathcal{F}, \mathbb{P})$.

However, for any partition $0 = t_0 < t_1 < \ldots < t_n = T$,

$$\sum_{j=0}^{n-1} \operatorname{sign}\left(\mathbb{E}\left(S^H_{t_{j+1}} - S^H_{t_j} \mid \mathcal{F}^{S^H}_{t_j}\right)\right) \mathbf{1}_{(t_j, t_{j+1}]} \in \beta(\mathcal{F}^{S^H}),$$

and

$$\begin{aligned}
&\left\| I_{S^H}\left(\sum_{j=0}^{n-1} \operatorname{sign}\left(\mathbb{E}\left(S^H_{t_{j+1}} - S^H_{t_j} \mid \mathcal{F}^{S^H}_{t_j}\right)\right) \mathbf{1}_{(t_j, t_{j+1}]}\right)\right\|_1 \\
&= \left\| \sum_{j=0}^{n-1} \operatorname{sign}\left(\mathbb{E}\left(S^H_{t_{j+1}} - S^H_{t_j} \mid \mathcal{F}^{S^H}_{t_j}\right)\right) \left(S^H_{t_{j+1}} - S^H_{t_j}\right)\right\|_1 \\
&= \mathbb{E}\left(\left| \sum_{j=0}^{n-1} \operatorname{sign}\left(\mathbb{E}\left(S^H_{t_{j+1}} - S^H_{t_j} \mid \mathcal{F}^{S^H}_{t_j}\right)\right) \left(S^H_{t_{j+1}} - S^H_{t_j}\right)\right|\right) \\
&\geq \mathbb{E}\left(\sum_{j=0}^{n-1} \operatorname{sign}\left(\mathbb{E}\left(S^H_{t_{j+1}} - S^H_{t_j} \mid \mathcal{F}^{S^H}_{t_j}\right)\right) \left(S^H_{t_{j+1}} - S^H_{t_j}\right)\right) \\
&= \sum_{j=0}^{n-1} \mathbb{E}\left(\operatorname{sign}\left(\mathbb{E}\left(S^H_{t_{j+1}} - S^H_{t_j} \mid \mathcal{F}^{S^H}_{t_j}\right)\right) \left(S^H_{t_{j+1}} - S^H_{t_j}\right)\right) \\
&= \sum_{j=0}^{n-1} \mathbb{E}\left(\mathbb{E}\left(\operatorname{sign}\left(\mathbb{E}\left(S^H_{t_{j+1}} - S^H_{t_j} \mid \mathcal{F}^{S^H}_{t_j}\right)\right) \left(S^H_{t_{j+1}} - S^H_{t_j}\right) \middle| \mathcal{F}^{S^H}_{t_j}\right)\right) \\
&= \sum_{j=0}^{n-1} \mathbb{E}\left(\operatorname{sign}\left(\mathbb{E}\left(S^H_{t_{j+1}} - S^H_{t_j} \mid \mathcal{F}^{S^H}_{t_j}\right)\right) \mathbb{E}\left(\left(S^H_{t_{j+1}} - S^H_{t_j}\right) \middle| \mathcal{F}^{S^H}_{t_j}\right)\right) \\
&= \sum_{j=0}^{n-1} \left\| \mathbb{E}\left(S^H_{t_{j+1}} - S^H_{t_j} \mid \mathcal{F}^{S^H}_{t_j}\right)\right\|_1 .
\end{aligned}$$

Then, S^H is a quasi-martingale. Hence, the proof of the lemma is complete. □

THEOREM 3.14.– *If, for all $1 \leq k \leq N$, we have that $H_k \geq 1/2$, and there exists $1 \leq k_0 \leq N$ such that $1/2 < H_{k_0} < 3/4$, then $S^H(N, a)$ is not a quasi-martingale, and in particular, it is not a semi-martingale.*

PROOF.– For $n \in \mathbb{N}$ and $j \in \{1, 2, \ldots, n\}$, let us denote

$$\Delta_j^n S^H = S^H_{\frac{Tj}{n}} - S^H_{\frac{T(j-1)}{n}}.$$

Because conditional expectation is a contraction with respect to $L^1(\Omega, \mathcal{F}, \mathbb{P})$ - norm, we have, for all $n \in \mathbb{N}$ and all $j = 1, \ldots, n-1$, the following inequalities:

$$\left\| \mathbb{E}\left(\Delta_{j+1}^n S^H | \Delta_j^n S^H\right) \right\|_1 \leq \left\| \mathbb{E}\left(\Delta_{j+1}^n S^H | \mathcal{F}^X_{\frac{Tj}{n}}\right) \right\|_1.$$

Moreover, because $\mathbb{E}\left(\Delta_{j+1}^n S^H | \Delta_j^n S^H\right)$ is a centered Gaussian random variable, we can proceed as follows:

$$\left\| \mathbb{E}\left(\Delta_{j+1}^n S^H | \Delta_j^n S^H\right) \right\|_1 = \sqrt{\frac{2}{\pi}} \left\| \mathbb{E}\left(\Delta_{j+1}^n S^H | \Delta_j^n S^H\right) \right\|_2.$$

Furthermore, applying proposition 1.1, we can proceed further as

$$\sum_{j=1}^{n-1} \left\| \mathbb{E}\left(\Delta_{j+1}^n S^H | \mathcal{F}^X_{\frac{Tj}{n}}\right) \right\|_1 \geq \sqrt{\frac{2}{\pi}} \sum_{j=1}^{n-1} \left\| \mathbb{E}\left(\Delta_{j+1}^n S^H | \Delta_j^n S^H\right) \right\|_2$$

$$= \sqrt{\frac{2}{\pi}} \sum_{j=1}^{n-1} \left\| \frac{\mathrm{Cov}\left(\Delta_{j+1}^n S^H, \Delta_j^n S^H\right)}{\mathbb{E}\left(\Delta_j^n S^H\right)^2} \Delta_j^n S^H \right\|_2$$

$$= \sqrt{\frac{2}{\pi}} \sum_{j=1}^{n-1} \frac{\mathrm{Cov}\left(\Delta_{j+1}^n S^H, \Delta_j^n S^H\right)}{\sqrt{\mathbb{E}\left(\Delta_j^n S^H\right)^2}}.$$

Now, consider the numerator and denominator separately. On the one hand, it follows from lemma 3.5 that

$$\mathrm{Cov}\left(\Delta_{j+1}^n S^H, \Delta_j^n S^H\right) = C\left(\frac{T(j-1)}{n}, \frac{Tj}{n}, \frac{Tj}{n}, \frac{T(j+1)}{n}\right)$$

$$= \sum_{i=1}^{N} \frac{a_i^2 T^{2H_i}}{2n^{2H_i}} \left(2^{2H_i}(2j^{2H_i} + 1) - 2 - (2j+1)^{2H_i} - (2j-1)^{2H_i}\right).$$

Furthermore, taking into account the inequality

$$2^{2H_i}(2j^{2H_i} + 1) - 2 - (2j+1)^{2H_i} - (2j-1)^{2H_i} \geq 0,$$

for which, from the proof of corollary 3.3, we can deduce that

$$\operatorname{Cov}\left(\Delta_{j+1}^n S^H, \Delta_j^n S^H\right) \geq \frac{a_{k_0}^2 T^{2H_{k_0}}}{2n^{2H_{k_0}}}\Big(2^{2H_{k_0}}(2j^{2H_{k_0}}+1) \\ -2-(2j+1)^{2H_{k_0}}-(2j-1)^{2H_{k_0}}\Big).$$

On the other hand, it follows from proposition 3.2 and the fact that $H_k \geq \frac{1}{2}$ for all $1 \leq k \leq N$ that

$$\mathbb{E}\left(\Delta_j^n S^H\right)^2 = \sum_{i=1}^{N} \frac{a_i^2 T^{2H_i}}{n^{2H_i}}\left(-2^{2H_i-1}(j^{2H_i}+(j-1)^{2H_i})+(2j-1)^{2H_i}+1\right) \\ \leq \frac{1}{n}\sum_{i=1}^{N} a_i^2 T^{2H_i}\left(-2^{2H_i-1}(j^{2H_i}+(j-1)^{2H_i})+(2j-1)^{2H_i}+1\right)$$

Therefore, using the notation

$$I_n = \sum_{j=1}^{n-1} \frac{\operatorname{Cov}\left(\Delta_{j+1}^n S^H, \Delta_j^n S^H\right)}{\sqrt{\mathbb{E}\left(\Delta_j^n S^H\right)^2}},$$

we obtain the following lower bound

$$I_n \geq \frac{a_{k_0}^2 T^{2H_{k_0}}}{2n^{2H_{k_0}-1/2}} \sum_{j=1}^{n-1} \frac{u_j(H_{k_0})}{v_j},$$

where, for any $j \in \mathbb{N}$,

$$u_j(H_{k_0}) = 2^{2H_{k_0}}(2j^{2H_{k_0}}+1)-2-(2j+1)^{2H_{k_0}}-(2j-1)^{2H_{k_0}},$$

and

$$v_j = \left(\sum_{i=1}^{N} a_i^2 T^{2H_i}\left(-2^{2H_i-1}(j^{2H_i}+(j-1)^{2H_i})+(2j-1)^{2H_i}+1\right)\right)^{1/2}.$$

On one hand, the terms in the numerator have the following asymptotic behavior:

$$
\begin{aligned}
u_n(H_{k_0}) &= 2^{2H_{k_0}}(2n^{2H_{k_0}}+1) - 2 - (2n+1)^{2H_{k_0}} - (2n-1)^{2H_{k_0}} \\
&= 2^{2H_{k_0}}(2n^{2H_{k_0}}+1) - 2 - 2^{2H_{k_0}}n^{2H_{k_0}}\left(1+\frac{1}{2n}\right)^{2H_{k_0}} \\
&\quad - 2^{2H_{k_0}}n^{2H_{k_0}}\left(1-\frac{1}{2n}\right)^{2H_{k_0}} = 2^{2H_{k_0}}(2n^{2H_{k_0}}+1) - 2 \\
&\quad - 2^{2H_{k_0}}n^{2H_{k_0}}\left(1+\frac{H_{k_0}}{n}+\frac{H_{k_0}(2H_{k_0}-1)}{4n^2}+o\left(\frac{1}{n^2}\right)\right) \\
&\quad - 2^{2H_{k_0}}n^{2H_{k_0}}\left(1-\frac{H_{k_0}}{n}+\frac{H_{k_0}(2H_{k_0}-1)}{4n^2}+o\left(\frac{1}{n^2}\right)\right) \\
&= 2^{2H_{k_0}}\left(1 - 2^{1-2H_{k_0}} - \frac{n^{2H_{k_0}-2}}{2}H_{k_0}(2H_{k_0}-1) + n^{2H_{k_0}-2}o\left(\frac{1}{n}\right)\right) \\
&= 2^{2H_{k_0}} - 2 - 2^{2H_{k_0}}\frac{H_{k_0}(2H_{k_0}-1)}{2n^{2-2H_{k_0}}} + o\left(\frac{1}{n^{2-2H_{k_0}}}\right),
\end{aligned}
$$

and consequently, taking into account that $H_{k_0} \in [\frac{1}{2}, 1)$, we get the limit relation

$$
\lim_{n\to+\infty} u_n(H_{k_0}) = 2^{2H_{k_0}} - 2.
$$

On the other hand, the denominator can be expanded as

$$
\begin{aligned}
v_n^2 &= \sum_{i=1}^{N} a_i^2 T^{2H_i}\Big(-2^{2H_i-1}(n^{2H_i}+(n-1)^{2H_i}) + (2n-1)^{2H_i}+1\Big) \\
&= \sum_{i=1}^{N} a_i^2 T^{2H_i}\left(-2^{2H_i-1}n^{2H_i}\left(1+\left(1-\frac{1}{n}\right)^{2H_i}\right) + 2^{2H_i}n^{2H_i}\left(1-\frac{1}{2n}\right)^{2H_i}+1\right) \\
&= \sum_{i=1}^{N} a_i^2 T^{2H_i}\left(-2^{2H_i-1}n^{2H_i}\left(2-\frac{2H_i}{n}+\frac{H_i(2H_i-1)}{n^2}+o\left(\frac{1}{n^2}\right)\right)\right. \\
&\quad \left. +2^{2H_i}n^{2H_i}\left(1-\frac{H_i}{n}+\frac{H_i(2H_i-1)}{4n^2}+o\left(\frac{1}{n^2}\right)\right)+1\right) \\
&= \sum_{i=1}^{N} a_i^2 T^{2H_i}\left(1 - 2^{2H_i}n^{2H_i-2}\frac{H_i(2H_i-1)}{4} + o\left(n^{2H_i-2}\right)\right),
\end{aligned}
$$

and consequently, because $H_i \in [\frac{1}{2}, 1)$ for all $1 \leq i \leq N$, we obtain that

$$\lim_{n \to +\infty} v_n = \left(\sum_{i=1}^{N} a_i^2 T^{2H_i} \right)^{1/2}.$$

Thus,

$$\lim_{n \to +\infty} \frac{u_n(H_{k_0})}{v_n} = \frac{2^{2H_{k_0}} - 2}{\left(\sum_{i=1}^{N} a_i^2 T^{2H_i} \right)^{1/2}}.$$

According to the Stolz–Cesàro theorem, the sequence of Cesàro means that $\left\{ \frac{1}{n} \sum_{j=1}^{n-1} \frac{u_j(H_{k_0})}{v_j}, n \geq 1 \right\}$ converges to the same limit as $n \to \infty$. That is,

$$\lim_{n \to +\infty} \frac{1}{n} \sum_{j=1}^{n-1} \frac{u_j(H_{k_0})}{v_j} = \frac{2^{2H_{k_0}} - 2}{\left(\sum_{i=1}^{N} a_i^2 T^{2H_i} \right)^{1/2}} > 0.$$

Because

$$I_n \geq \frac{a_{k_0}^2 T^{2H_{k_0}}}{2n^{2H_{k_0}-1/2}} \sum_{j=1}^{n-1} \frac{u_j(H_{k_0})}{v_j} = \frac{a_{k_0}^2 T^{2H_{k_0}}}{2n^{2H_{k_0}-3/2}} \frac{1}{n} \sum_{j=1}^{n-1} \frac{u_j(H_{k_0})}{v_j},$$

and because $1/2 < H_{k_0} < 3/4$, $\lim_{n \to \infty} \frac{a_{k_0}^2 T^{2H_{k_0}}}{2n^{2H_{k_0}-\frac{3}{2}}} = +\infty$.

Consequently, $\lim_{n \to \infty} I_n = +\infty$, which yields

$$\sup_{\pi} \sum_{j=0}^{k-1} \left\| E\left(X_{t_{j+1}} - X_{t_j} | \mathcal{F}_{t_j}^X \right) \right\|_1 = \infty. \qquad \square$$

THEOREM 3.15.– *If, for all $1 \leq i \leq N$, we have that $H_i \in (1/2, 1)$ and one of the Hurst indices, say, H_{k_0}, equals $3/4$, then the process $S^H(a)$ is not a quasi-martingale.*

PROOF.– Because conditional expectation is a contraction with respect to $L^1(\Omega, \mathcal{F}, \mathbb{P})$ - norm, we have, for all $n \in \mathbb{N}$ and all $j = 1, \ldots, n-1$, the following relations:

$$\left\| \mathbb{E}\left(\Delta_{j+1}^n S^H | \Delta_j^n S^H, \ldots, \Delta_1^n S^H \right) \right\|_1 \leq \left\| \mathbb{E}\left(\Delta_{j+1}^n S^H | \mathcal{F}_{\frac{Tj}{n}}^{S^H} \right) \right\|_1.$$

Moreover, because $\mathbb{E}\left(\Delta_{j+1}^n S^H | \Delta_j^n S^H, \ldots, \Delta_1^n S^H\right)$ is a centered Gaussian random variable,

$$\left\|\mathbb{E}\left(\Delta_{j+1}^n S^H | \Delta_j^n S^H, \ldots, \Delta_1^n S^H\right)\right\|_1 = \sqrt{\frac{2}{\pi}} \left\|\mathbb{E}\left(\Delta_{j+1}^n S^H | \Delta_j^n S^H, \ldots, \Delta_1^n S^H\right)\right\|_2.$$

Consequently,

$$\sum_{j=1}^{n-1} \left\|\mathbb{E}\left(\Delta_{j+1}^n S^H | \mathcal{F}_{\frac{j}{n}}^{S^H}\right)\right\|_1 \geq \sqrt{\frac{2}{\pi}} \sum_{j=1}^{n-1} \left\|\mathbb{E}\left(\Delta_{j+1}^n S^H | \Delta_j^n S^H, \ldots, \Delta_1^n S^H\right)\right\|_2,$$

and the lemma will be proved if we show that

$$\lim_{n\to\infty} \sum_{j=1}^{n-1} \left\|\mathbb{E}\left(\Delta_{j+1}^n S^H | \Delta_j^n S^H, \ldots, \Delta_1^n S^H\right)\right\|_2 = \infty.$$

For $n \in \mathbb{N}$ and $j = 1, \ldots, n-1$,

$$\left(\Delta_{j+1}^n S^H, \Delta_j^n S^H, \ldots, \Delta_1^n S^H\right)$$

is a Gaussian vector, and from corollary 3.7, the covariance matrix of the vector $\left(\Delta_j^n S^H, \ldots, \Delta_1^n S^H\right)$ is invertible. Therefore, by virtue of corollary 1.4, we have

$$\mathbb{E}\left(\Delta_{j+1}^n S^H | \Delta_j^n S^H, \ldots, \Delta_1^n S^H\right) = \sum_{k=1}^{j} b_k \Delta_k^n S^H, \qquad [3.54]$$

where the vector $b = (b_1, \ldots, b_j)^T$ equals

$$b = A^{-1}(j) m(j), \qquad [3.55]$$

$m(j)$ is a j-dimensional vector whose kth component $m_k(j)$ equals

$$Cov\left(\Delta_{j+1}^n S^H, \Delta_k^n S^H\right)$$

and $A(j)$ is the covariance matrix of the Gaussian vector

$$\left(\Delta_1^n S^H, \ldots, \Delta_j^n S^H\right).$$

Note that $A(j)$ is symmetric and, because the random variables

$$\Delta_1^n S^H, \dots, \Delta_j^n S^H$$

are linearly independent, $A(j)$ is also positive definite. It follows from equations [3.54] and [3.55] that

$$\left\|\mathbb{E}\left(\Delta_{j+1}^n S^H | \Delta_j^n S^H, \dots, \Delta_1^n S^H\right)\right\|_2^2 = b^T A(j) b = (m(j))^T A^{-1}(j) m(j). \quad [3.56]$$

Now, by using theorem A.12 from the Appendix and equation [3.56], we get

$$\begin{aligned}
&\left\|\mathbb{E}\left(\Delta_{j+1}^n S^H | \Delta_j^n S^H, \dots, \Delta_1^n S^H\right)\right\|_2^2 \\
&= \frac{(m(j))^T A^{-1}(j) m(j)}{(m(j))^T m(j)} (m(j))^T m(j) \\
&\geq \min_{x \neq 0} \frac{x^T A^{-1}(j) x}{x^T x} (m(j))^T m(j) \geq (\lambda(j))^{-1} (m(j))^T m(j) \\
&= \|m(j)\|_2^2 (\lambda(j))^{-1},
\end{aligned}$$

where $(\lambda(j))^{-1}$ is the smallest eigenvalue of the matrix $(A(j))^{-1}$. In turn, this means that $\lambda(j)$ is the largest eigenvalue of the matrix $A(j)$. Set, for $l \in \{1, \dots, N\}$,

$$C_l = (C_{i,k}^l)_{1 \leq i,k \leq j}$$

the covariance matrix of the increments of the sfBm ξ^{H_l}. We have

$$A(j) = \sum_{l=1}^N a_l^2 \, C_l,$$

and consequently, it follows from corollary A.6 in the Appendix that

$$\lambda(j) \leq \sum_{l=1}^N a_l^2 \, \mu_l, \quad [3.57]$$

where μ_l is the largest eigenvalue of the matrix C_l. From lemma 3.5, we also have

$$\begin{aligned}
C_{ik}^l = \frac{T^{2H_l}}{2n^{2H_l}} \big((| k - i | + 1)^{2H_l} - 2 | k - i |^{2H_l} + || k - i | - 1 |^{2H_l} \\
+ 2(k + i - 1)^{2H_l} - (k + i)^{2H_l} - (k + i - 2)^{2H_l} \big) = \frac{T^{2H_l}}{2n^{2H_l}} (E_{ik}^l + F_{ik}^l),
\end{aligned}$$

where

$$E_{ik}^{l}=2\left(\frac{\left(|k-i|+1\right)^{2H_l}+||\,k-i\,|-1\,|^{2H_l}}{2}-|\,k-i\,|^{2H_l}\right)$$

and

$$F_{ik}^{l}=2\left((k+i-1)^{2H_l}-\frac{(k+i)^{2H_l}+(k+i-2)^{2H_l}}{2}\right).$$

Note that the convexity of the function $x\rightarrow x^{2H_l}$, $x\geq 0$, implies that $E_{ik}^{l}\geq 0$ and $F_{ik}^{l}\leq 0$. Moreover, because $H_l\geq 1/2$, corollary 3.3 yields that $C_{ik}^{l}\geq 0$.

Therefore, using theorem A.10 from the Appendix, we obtain

$$\mu_l\leq\max_{i=1,\ldots,j}\sum_{i=1}^{j}|\,C_{ik}^{l}\,|\leq\frac{T^{2H_l}}{2n^{2H_l}}\max_{i=1,..j}\sum_{k=1}^{j}E_{ik}^{l},$$

and consequently,

$$\begin{aligned}\mu_l&\leq\frac{T^{2H_l}}{n^{2H_l}}\sum_{k=1}^{j}\left(2\left(\frac{\left(|\,k-1\,|+1\right)^{2H_l}+||\,k-1\,|-1\,|^{2H_l}}{2}-|\,k-1\,|^{2H_l}\right)\right.\\&=\frac{T^{2H_l}}{n^{2H_l}}\sum_{k=1}^{j}\left(\left(|\,k-1\,|+1\right)^{2H_l}+||\,k-1\,|-1\,|^{2H_l}-2\,|\,k-1\,|^{2H_l}\right)\\&=\frac{T^{2H_l}}{n^{2H_l}}\sum_{k'=0}^{j-1}\left((k'+1)^{2H_l}-2\,k'^{2H_l}+|\,k'-1\,|^{2H_l}\right)\\&=\frac{T^{2H_l}}{n^{2H_l}}\left(1+j^{2H_l}-(j-1)^{2H_l}\right)\\&\leq T^{2H_l}\left(\frac{1}{n^{2H_l}}+\frac{1}{n^{2H_l}}\max_{j-1\leq x\leq j}\frac{d(x^{2H_l})}{dx}\right)\\&\leq\frac{(1+2H_l)}{2n}T^{2H_l}.\end{aligned}$$

Hence, combining equality [3.57] with the above result, we obtain

$$\lambda^{-1}(j)\geq\alpha n, \qquad [3.58]$$

where $\alpha = \left(\sum_{l=1}^{N}(1+2H_l)T^{2H_l}\right)^{-1}$.

Next, let us determine a suitable lower bound of $\|m(j)\|_2^2$. From lemma 3.5, we have

$$\begin{aligned}\|m(j)\|_2^2 &= \sum_{k=1}^{j}\left(\mathrm{Cov}\left(\Delta_{j+1}^{n}S^{H_l},\Delta_k^n S^{H_l}\right)\right)^2\\ &= \sum_{k=1}^{j}\left(\sum_{l=1}^{N}\frac{T^{2H_l}a_l^2}{2n^{2H_l}}(f_{1,l}(k)-f_{2,l}(k))\right)^2,\end{aligned} \quad [3.59]$$

where

$$f_{1,l}(k) = (j-k+2)^{2H_l} - 2(j-k+1)^{2H_l} + (j-k)^{2H_l}$$

and

$$f_{2,l}(k) = (j+k+1)^{2H_l} - 2(j+k)^{2H_l} + (j+k-1)^{2H_l}.$$

From lemma A.4 in the Appendix, for any $1 \le l \le N$ and $1 \le k \le j$, we have that $f_{1,l}(k) \ge 0$ and $f_{2,l}(k) \ge 0$, $f_{1,l}(k) - f_{2,l}(k) > 0$ and

$$f_{1,l}(k)-f_{2,l}(k) \ge 2H_l(2H_l-1)\left((j-k+1)^{2H_l-2}-(j+k+1)^{2H_l-2}\right) \ge 0. \quad [3.60]$$

Combining equation [3.59] with equation [3.60] and because $H_{k_0} = 3/4$, we get

$$\begin{aligned}\|m(j)\|_2^2 &\ge \sum_{k=1}^{j}\left(\sum_{l=1}^{N}\frac{T^{2H_l}a_l^2}{n^{2H_l}}\,H_l(2H_l-1)\right.\\ &\qquad\left.\times\left((j-k+1)^{2H_l-2}-(j+k+1)^{2H_l-2}\right)\right)^2\\ &\ge \sum_{k=1}^{j}\left(\frac{T^{2H_{k_0}}a_{k_0}^2}{n^{2H_{k_0}}}\,H_{k_0}(2H_{k_0}-1)\right.\\ &\qquad\left.\times\left((j-k+1)^{2H_{k_0}-2}-(j+k+1)^{2H_{k_0}-2}\right)\right)^2\\ &= \frac{9}{64}\,\frac{T^3a_{k_0}^4}{n^3}\sum_{k=1}^{j}\left((j-k+1)^{-1/2}-(j+k+1)^{-1/2}\right)^2.\end{aligned} \quad [3.61]$$

For any integer $j \geq 1$, let us consider the function

$$f_j(x) = (j - x + 1)^{-1/2} - (j + x - 1)^{-1/2}, \ 1 \leq x \leq j.$$

As f_j increases, we have

$$\sum_{k=2}^{j} \left((j - k + 1)^{-1/2} - (j + k - 1)^{-1/2} \right)^2 \geq \int_1^j f_j(x)^2 \, dx. \qquad [3.62]$$

But

$$\begin{aligned} \int_1^j f_j(x)^2 \, dx &= \int_1^j \left(\frac{1}{j - x + 1} + \frac{1}{j + x - 1} - 2 \frac{1}{\sqrt{j^2 - (x-1)^2}} \right) dx \\ &= \ln(2j - 1) - 2 \int_1^j \frac{1}{\sqrt{j^2 - (x-1)^2}} \, dx \qquad [3.63] \\ &= \ln(2j - 1) + 2 \arccos\left(\frac{j-1}{j} \right) - \pi. \end{aligned}$$

Hence, combining equation [3.58] with equations [3.61], [3.62] and [3.63], we get

$$\|m(j)\|_2^2 \, \lambda^{-1}(j) \geq \frac{\beta}{n^2} \left(\ln(2j - 1) + 2 \arccos\left(\frac{j-1}{j} \right) - \pi \right), \qquad [3.64]$$

where $\beta = \dfrac{9}{64} T^3 a_{k_0}^4 \alpha$, where α is the constant defined in equation [3.58].

Combining equation [3.64] with equation [3.56], we obtain that

$$\begin{aligned} &\sum_{j=1}^{n-1} \|E\left(\Delta_{j+1}^n S_{3/4} \mid \Delta_j^n S_{3/4}, .., \Delta_1^n S_{3/4} \right)\|_2 \\ &\geq \frac{\sqrt{\beta}}{n} \sum_{j=1}^{n-1} \sqrt{\ln(2j - 1) + 2 \arccos\left(\frac{j-1}{j} \right) - \pi}. \end{aligned}$$

Because $\lim_{n \to \infty} \sqrt{\ln(2n - 1) + 2 \arccos\left(\frac{n-1}{n} \right) - \pi} = \infty$, using the Césaro theorem, we have

$$\lim_{n \to \infty} \frac{\sqrt{\beta}}{n} \sum_{j=1}^{n-1} \sqrt{\ln(2j - 1) + 2 \arccos\left(\frac{j-1}{j} \right) - \pi} = \infty,$$

which completes the proof. □

3.7. Mixed sub-fractional colored-white heat equation

In this section, we introduce a stochastic heat equation with a colored-white fractional noise, which behaves as a Wiener process in space variable and as a mixed sub-fractional Brownian motion in time. We give a sufficient condition for the existence of its mild solution and regularity analyze the properties of this solution with respect to the time variable. We also determine the Hausdorff dimensions of the graphs and ranges of its sample paths.

Let $S = \{S_t^H(N,a),\ t \geq 0\} = \{S_t^H,\ t \geq 0\}$ be a mixed sub-fractional Brownian motion defined on a probability space $(\Omega, \mathcal{F}, \mathbb{P})$, where $N \in \mathbb{N}$, $H = (H_1, H_2, \ldots, H_N) \in (\frac{1}{2}, 1)^N$, and vector $a = (a_1, a_2, \ldots, a_N) \in \mathbb{R}^N, a_i \neq 0$, see definition 3.1.

As before, we assume that all H_i are different and that $\frac{1}{2} < H_1 < H_2 < \ldots < H_N$.

3.7.1. *Moving average representation of msfBm*

In the following lemma, we give a moving average representation of the msfBm.

LEMMA 3.14.– *For any t,*

$$S_t^H(a) = \sum_{i=1}^{N} C_1(H_i)\left(H_i - \frac{1}{2}\right)\int_{\mathbb{R}}\int_0^t \left((u-s)_+^{H_i-\frac{3}{2}} + (u+s)_-^{H_i-\frac{3}{2}}\right)_d u dW_i(s), \qquad [3.65]$$

where $W_i, 1 \leq i \leq N$ are independent Wiener processes on $\mathbb{R}$, and

$$C_1(H_i) = \left(\int_0^\infty \left((1+s)^{H_i-\frac{1}{2}} - s^{H_i-\frac{1}{2}}\right)^2 ds + \frac{1}{2H_i}\right)^{-\frac{1}{2}} = \frac{(2H_i \sin(\pi H_i)\Gamma(2H_i))^{1/2}}{\Gamma(H_i + 1/2)}. \qquad [3.66]$$

PROOF.– For fixed $1 \leq i \leq N$, from corollary 2.2, we have the following representation of sfBm:

$$\xi_t^{H_i} = C_1(H_i)\int_{\mathbb{R}}\left[(t-s)_+^{H_i-\frac{1}{2}} + (t+s)_-^{H_i-\frac{1}{2}} - 2(-s)_+^{H_i-\frac{1}{2}}\right] dW_i(s). \qquad [3.67]$$

The independence of processes $\xi^{H_i}, i \in \{1, \dots, N\}$ and equality [3.67] imply that

$$S_t^H(a) = \sum_{i=1}^{N} C_1(H_i) \int_{\mathbb{R}} \left((t-s)_+^{H_i - \frac{1}{2}} + (t+s)_-^{H_i - \frac{1}{2}} - 2(-s)_+^{H_i - \frac{1}{2}} \right) dW_i(s), \quad [3.68]$$

where W_i $1 \leq i \leq N$ are independent Wiener processes on $\mathbb{R}$, and $C_1(H_i)$ is defined by equation [3.66]. Then, we easily check that

$$\begin{aligned} &(t-s)_+^{H_i - \frac{1}{2}} + (t+s)_-^{H_i - \frac{1}{2}} - 2(-s)_+^{H_i - \frac{1}{2}} \\ &= \left(H_i - \frac{1}{2} \right) \int_0^t \left((u-s)_+^{H_i - \frac{3}{2}} + (u+s)_-^{H_i - \frac{3}{2}} \right) du, \end{aligned} \quad [3.69]$$

for all $H_i \in \left(\frac{1}{2}, 1 \right)$ and $(t, s) \in \mathbb{R}^2$. Finally, from equations [3.68] and [3.69], we deduce equality [3.65]. □

The following proposition will play an important role throughout this section.

PROPOSITION 3.5.– *We have*

$$\frac{\partial^2 \text{Cov}(S_u^H, S_v^H)}{\partial u \partial v} = \sum_{i=1}^{N} a_i^2 H_i (2H_i - 1) \Big(\mid u - v \mid^{2H_i - 2} - (u+v)^{2H_i - 2} \Big). \quad [3.70]$$

Moreover, there exists a positive constant D such that

$$\frac{\partial^2 \text{Cov}(S_u^H, S_v^H)}{\partial u \partial v} \leq D |u - v|^{2H_1 - 2} \quad [3.71]$$

for any $u, v \in [0, T]$.

PROOF.– Relation [3.70] is a straightforward consequence of equality [3.3]. Therefore, let us prove equation [3.71]. It follows from relation [3.70] and the evident inequality

$$0 \leq |u - v|^{2H_i - 2} - (u+v)^{2H_i - 2} \leq |u - v|^{2H_i - 2}$$

that

$$\begin{aligned} \frac{\partial^2 \text{Cov}(S_u^H, S_v^H)}{\partial u \partial v} &\leq \sum_{i=1}^{N} a_i^2 H_i (2H_i - 1) |u - v|^{2H_i - 2} \\ &\leq |u - v|^{2H_1 - 2} \sum_{i=1}^{N} a_i^2 H_i (2H_i - 1) |u - v|^{2(H_i - H_1)} \leq D |u - v|^{2H_1 - 2} \end{aligned}$$

with $D = 2 \sum_{i=1}^{N} a_i^2 H_i (2H_i - 1) T^{2(H_i - H_1)}$. □

3.7.2. *The heat equation driven by mixed sub-fractional noise*

In this section, we study the stochastic partial differential equation

$$\begin{cases} \dfrac{\partial u_{a,H}}{\partial t} = \dfrac{1}{2}\Delta u_{a,H} + W_{a,H}, \; t \in [0,T], \; x \in \mathbb{R}^d, \\ u_{a,H}(\cdot, 0) = 0, \end{cases} \qquad [3.72]$$

where Δ is the Laplacian operator $\Delta = \sum_{j=1}^{d} \frac{\partial^2}{\partial x_j}$, and

$$W_{a,H} = \{W_{a,H}(t, A) | t \in [0,T], A \in \mathcal{B}_b(\mathbb{R}^d)\}$$

is a centered random noise with covariance function given by the equality

$$\mathbb{E}(W_{a,H}(t, A) W_{a,H}(s, B)) = \mathrm{Cov}(S_t^H, S_s^H)\lambda(A \cap B),$$

where λ is the Lebesgue measure and $\mathcal{B}_b(\mathbb{R}^d)$ denotes the class of bounded Borel sets in $\mathbb{R}^d$. We call this noise mixed sub-fractional-white because it behaves as an msfBm in time and as a Wiener process (white) in space. The stochastic partial differential equation [3.72] will be called the mixed sub-fractional heat equation.

The canonical Hilbert space associated with the noise $W_{a,H}$ is defined as follows. First, consider the set $\mathcal{E}$ of linear combinations of elementary functions $\mathbf{1}_{[0,t]} \times A$, $t \in [0,T]$, $A \in \mathcal{B}_b(\mathbb{R}^d)$, and $\mathcal{H}_{a,H}$ be the Hilbert space defined as the closure of $\mathcal{E}$ with respect to the inner product

$$\langle \mathbf{1}_{[0,t]} \times A, \mathbf{1}_{[0,s]} \times B \rangle_{\mathcal{H}_{a,H}} := \mathbb{E}(W_{a,H}(t, A) W_{a,H}(s, B)).$$

We have for smooth enough $g, h \in \mathcal{H}_{a,H}$ that

$$\langle g, h \rangle_{\mathcal{H}_{a,H}} = \int_0^T \int_0^T \int_{\mathbb{R}^d} \frac{\partial^2 \mathrm{Cov}(S_u^H, S_v^H)}{\partial u \partial v} g(y,u) h(y,v) dy dv du.$$

With a routine extension of the construction described, for example, in [TUD 13] and [TUD 14], it is possible to define the Wiener integral with respect to the process $W_{a,H}$. This Wiener integral will act as an isometry between the Hilbert space $\mathcal{H}_{a,H}$ and $L^2(\Omega, \mathcal{F}, \mathbb{P})$ via the following equality

$$\mathbb{E}\left(\int_0^T \int_{\mathbb{R}^d} \varphi(u,y) W_{a,H}(du, dy) \int_0^T \int_{\mathbb{R}^d} \psi(u,y) W_{a,H}(du, dy) \right)$$
$$= \int_0^T \int_0^T \int_{\mathbb{R}^d} \varphi(u,y) \psi(v,y) dy \mu_{a,H}(du, dv) \qquad [3.73]$$

for any function φ, ψ such that

$$\int_0^T \int_0^T \int_{\mathbb{R}^d} |\varphi(u,y)|\, |\varphi(v,y)| dy\, d|\mu_{a,H}|(u,v) < \infty$$

and

$$\int_0^T \int_0^T \int_{\mathbb{R}^d} |\psi(u,y)|\, |\psi(v,y)| dy\, d|\mu_{a,H}|(u,v) < \infty,$$

where $\mu_{a,H}$ is the measure defined by

$$d\mu_{a,H}(u,v) = \frac{\partial^2 \mathrm{Cov}(S_u^H, S_v^H)}{\partial u \partial v} dudv \qquad [3.74]$$

and $|\mu_{a,H}|$ denotes the total variation measure associated with $\mu_{a,H}$ (see definition A.23 in the Appendix). In the case under consideration,

$$d|\mu_{a,H}|(u,v) = \left| \frac{\partial^2 \mathrm{Cov}(S_u^H, S_v^H)}{\partial u \partial v} \right| dudv.$$

The following formula will be useful later.

PROPOSITION 3.6.– *For any $g \in \mathcal{H}_{a,H}$, we have*

$$\int_0^T \int_{\mathbb{R}^d} g(s,y) W_a^H(ds,dy) = \sum_{i=1}^N a_i C_1(H_i)(H_i - 1/2)$$
$$\times \int_{\mathbb{R}} \int_{\mathbb{R}^d} \int_{\mathbb{R}} \mathbf{1}_{(0,T)}(u) g(u,y) \left[(u-s)_+^{H_i - \frac{3}{2}} + (u+s)_-^{H_i - \frac{3}{2}} \right] du W_i(ds,dy). \qquad [3.75]$$

where $W_i, 1 \le i \le N$, are independent space-time white noises with covariance functions

$$\mathbb{E}(W_i(s,A)W_i(t,B)) = (t \wedge s)\lambda(A \cap B),$$

and $C(H_i)$ are defined by equation [3.66].

PROOF.– Proposition 3.6 is a straightforward consequence of the moving average representation n of the msfBm equation [3.68]. □

Now we will define the mild solution of the mixed sub-fractional heat equation [3.72].

DEFINITION 3.10.– *If we denote by* $\{u_{a,H}(t,x); t \in [0,T], x \in \mathbb{R}^d\}$*, the process defined by*

$$u_{a,H}(t,x) := \int_0^t \int_{\mathbb{R}^d} G(t-v, x-y) W_{a,H}(dv, dy), \qquad [3.76]$$

where the above integral is a Wiener integral with respect to the noise $W_{a,H}$ *and* G *is the Green kernel of the heat equation given by*

$$G(t,x) = \begin{cases} (2\pi t)^{-d/2} \exp\Big(-\frac{|x|^2}{2t}\Big) \text{ if } t > 0, x \in \mathbb{R}^d \\ 0 \text{ if } t \leq 0, x \in \mathbb{R}^d, \end{cases}$$

then the process $u_{a,H}$ *is called the mild solution of the stochastic heat equation [3.72].*

In order to see the link between formula [3.76] and equation [3.72], let us first recall that the heat equation is the partial differential equation

$$\frac{\partial u}{\partial t} = \frac{1}{2}\Delta u, \quad t \in [0,T], \quad x \in \mathbb{R}^d, \qquad [3.77]$$

where Δ is the Laplacian operator. Given any bounded continuous initial condition $u_0 : \mathbb{R}^d \longrightarrow \mathbb{R}$, there exists a unique solution u to equation [3.77], which is continuous on $[0,T] \times \mathbb{R}^d$ and such that $u(0,x) = u_0(x)$ for any $x \in \mathbb{R}^d$. This solution is given by

$$u(t,x) = \int_{\mathbb{R}^d} G(t, x-y) u_0(y) dy.$$

Let us now consider equation [3.77] with additional forcing term f:

$$\frac{\partial u}{\partial t} = \frac{1}{2}\Delta u + f, \quad t \in [0,T], \quad x \in \mathbb{R}^d. \qquad [3.78]$$

From the variations of constants formula, we obtain that the solution to equation [3.78] is given by

$$u(t,x) := \int_{\mathbb{R}^d} G(t, x-y) u_0(y) dy + \int_0^t \Big(\int_{\mathbb{R}^d} G(t-s, x-y) f(s,y) dy\Big) ds$$

In particular, in the simplest case $u_0 := 0$, the solution is given by the following formula

$$u(t,x) := \int_0^t \int_{\mathbb{R}^d} G(t-s, x-y) f(s,y) dy ds. \qquad [3.79]$$

Now, replacing f by the mixed sub-fractional white noise $W_{a,H}$, in the spirit of formula [3.79] construction, a natural candidate solution of equation [3.72] is the stochastic process defined by equation [3.76]. Further details on the mild solution can be found, for example, in [BAL 08] and [DAL 99].

The following lemma is useful in the study of the properties of the mild solution [3.76].

LEMMA 3.15.– *For any $x \in \mathbb{R}^d$, $0 < s \leq t$, $0 \leq u < t$ and $0 \leq v < s$, we have*

$$\int_{\mathbb{R}^d} G(t-u, x-y)G(s-v, x-y)dy = \left(\frac{1}{2\pi(t+s-u-v)}\right)^{d/2}. \quad [3.80]$$

PROOF.– By using the explicit representation of G, we get

$$\int_{\mathbb{R}^d} G(t-u, x-y)G(s-v, x-y)dy = \int_{\mathbb{R}^d} (2\pi(t-u))^{-d/2} \exp\Big(-\frac{|y-x|^2}{2(t-u)}\Big)$$
$$\times (2\pi(s-v))^{-d/2} \exp\Big(-\frac{|y-x|^2}{2(s-v)}\Big)dy = \int_{\mathbb{R}^d} K_1 \exp\Big(-K_2\,|y-x|^2\Big)dy,$$

where

$$K_1 = (2\pi(t-u))^{-d/2}(2\pi(s-v))^{-d/2} \text{ and } K_2 = \frac{1}{2(s-v)} + \frac{1}{2(t-u)}.$$

The known formula of the d-dimensional Gaussian integral is given by

$$\int_{\mathbb{R}^d} \exp\Big(-\alpha\,|y|^2\Big)dy = \Big(\frac{\pi}{\alpha}\Big)^{d/2},$$

for any $\alpha > 0$. Applying this with $\alpha = K_2$, we get the proof. □

3.7.3. *Existence of the solution*

In the following proposition, we give a sufficient condition for the existence of the mild solution to the mixed sub-fractional heat equation.

PROPOSITION 3.7.– *If $d < 4H_1$, then the solution to the mixed sub-fractional heat equation [3.72] exists.*

PROOF.– It follows from equations [3.73] and [3.76] that

$$\mathbb{E}\Big(u_{a,H}(t,x)^2\Big) = \int_0^t \int_0^t \int_{\mathbb{R}^d} G(t-u, x-y)G(t-v, x-y)dy\; \mu_{a,H}(du, dv). [3.81]$$

Therefore, it follows from equations [3.81] and [3.74] and proposition 3.5 that

$$\mathbb{E}\Big(u_{a,H}(t,x)^2\Big) \leq D\, I(t,x), \qquad [3.82]$$

where

$$I(t,x) = \int_0^t \int_0^t |u-v|^{2H_1-2} \int_{\mathbb{R}^d} G(t-u, x-y)G(t-v, x-y)dydudv. \qquad [3.83]$$

From equation [3.80], we get that

$$\int_{\mathbb{R}^d} G(t-u, x-y)G(t-v, x-y)dy = \left(\frac{1}{2\pi(2t-u-v)}\right)^{d/2}. \qquad [3.84]$$

Hence, from equations [3.82], [3.83] and [3.84], we deduce that if

$$\int_0^t \int_0^t |u-v|^{2H_1-2}(2t-u-v)^{-d/2}dudv < \infty, \qquad [3.85]$$

then the solution to the mixed sub-fractional heat equation [3.72] exists. To prove equation [3.85], let us write

$$\begin{aligned}
I &= \int_0^t \int_0^t |u-v|^{2H_1-2}(2t-u-v)^{-d/2}dudv \\
&= \int_0^t \int_0^u |u-v|^{2H_1-2}(2t-u-v)^{-d/2}dvdu \\
&\quad + \int_0^t \int_u^t |u-v|^{2H_1-2}(2t-u-v)^{-d/2}dvdu \\
&= I_1 + I_2.
\end{aligned}$$

Making the change of variables $x = u - v$, $y = 2t - u - v$, we get

$$I_1 = \frac{1}{2}\int\int_D x^{2H_1-2}y^{-d/2}dxdy,$$

where $D = \{(x,y)|\ 0 \leq x \leq 2t - y \text{ and } x \leq y \leq x + 2t\}$. We clearly have

$$D \subset \{(x,y)|0 \leq x \leq t \text{ and } x \leq y \leq x + 2t\},$$

which allows us to write

$$
\begin{aligned}
I_1 &\le \frac{1}{2}\int_0^t x^{2H_1-2}\int_x^{x+2t} y^{-d/2}dydx \\
&= \frac{1}{2-d}\int_0^t x^{2H_1-2}\Big((x+2t)^{1-d/2} - x^{1-d/2}\Big)dx \\
&= \frac{1}{2-d}\Big(\int_0^t x^{2H_1-2}(x+2t)^{1-d/2}dx - \int_0^t x^{2H_1-1-d/2}dx\Big).
\end{aligned}
$$

Because $H_1 > \frac{1}{2}$, the integral $\int_0^t x^{2H_1-2}(x+2t)^{1-d/2}dx$ is finite, and because of the assumption $d < 4H_1$, we have

$$
\int_0^t x^{2H_1-1-d/2}dx = \frac{1}{2H_1 - d/2}t^{2H_1-d/2} < \infty.
$$

Thus, $I_1 < \infty$. Applying the same technique, we prove that $I_2 < \infty$, and we deduce that $I < \infty$. □

3.7.4. *Covariance function and self-similarity of mild solution of mixed sub-fractional heat equation*

Throughout this section, we assume that

$$
d < 4H_1. \qquad [3.86]
$$

In the following proposition, we give an explicit expression of the covariance of the mild solution.

PROPOSITION 3.8.– *For fixed $x \in \mathbb{R}^d$ and for $s \le t$,*

$$
\begin{aligned}
&\mathbb{E}\Big(u_{a,H}(t,x)u_{a,H}(s,x)\Big) \\
&\quad = (2\pi)^{-d/2}\int_0^t\int_0^s \frac{\partial^2 \mathrm{Cov}(S_u^H, S_v^H)}{\partial u \partial v}(t+s-u-v)^{-\frac{d}{2}}dudv.
\end{aligned}
$$

PROOF.– It follows from equations [3.73] and [3.76] that

$$\begin{aligned}
&\mathbb{E}\Big(u_{a,H}(t,x)u_{a,H}(s,x)\Big) \\
&= \mathbb{E}\Bigg(\int_0^T \int_{\mathbb{R}^d} \mathbf{1}_{[0,t]}(v)G(t-v,x-y)W_{a,H}(dv,dy) \\
&\times \int_0^T \int_{\mathbb{R}^d} \mathbf{1}_{[0,s]}(v)G(s-v,x-y)W_{a,H}(dv,dy)\Bigg) \\
&= \int_0^T \int_0^T \int_{\mathbb{R}^d} \varphi(u,y)\psi(v,y)dy\, \mu_{a,H}(du,dv)
\end{aligned}$$

with

$$\varphi(u,y) := \mathbf{1}_{[0,t]}(u)G(t-u,x-y) \text{ and } \varphi(v,y) := \mathbf{1}_{[0,s]}(v)G(s-v,x-y).$$

This, together with equation [3.74], allows us to obtain the following relations:

$$\begin{aligned}
&\mathbb{E}\Big(u_{a,H}(t,x)u_{a,H}(s,x)\Big) = \int_0^t \int_0^s \frac{\partial^2 \mathrm{Cov}(S_u^H, S_v^H)}{\partial u \partial v} \\
&\times \int_{\mathbb{R}^d} G(t-u,x-y)G(s-v,x-y)dydudv.
\end{aligned}$$

Using equation [3.80], we get our statement. □

From equation [3.70] and proposition 3.8, we immediately get the following result.

COROLLARY 3.11.– *For fixed $x \in \mathbb{R}^d$ and for $(s,t) \in [0,T]^2$,*

$$\begin{aligned}
&\mathbb{E}\Big(u_{a,H}(t,x)u_{a,H}(s,x)\Big) = (2\pi)^{-d/2} \sum_{i=1}^N a_i^2 H_i(2H_i-1) \\
&\times \int_0^t \int_0^s \Big(|u-v|^{2H_i-2} - (u+v)^{2H_i-2}\Big)(t+s-u-v)^{-\frac{d}{2}} dudv.
\end{aligned}$$

The following proposition deals with the mixed self-similarity (see [ZIL 06]) of the mild solution's sample paths.

PROPOSITION 3.9.– *The process $u_{a,H} : t \longmapsto u_{a,H}(t,x)$ is mixed self-similar of order $H - \frac{d}{4}$. That is, for any $h > 0$, the processes $(u_{a,H}(ht,x))_{t\in\mathbb{R}_+}$ and $(u_{ah^{H-\frac{d}{4}},H}(t,x))_{t\in\mathbb{R}_+}$ have the same law, where*

$$ah^{H-\frac{d}{4}} = (a_1 h^{H_1-\frac{d}{4}}, \ldots, a_N h^{H_N-\frac{d}{4}}).$$

PROOF.– For any fixed $h > 0$, the processes $(u_{a,H}(ht,x))_{t\in\mathbb{R}_+}$ and $(u_{ah^{H-\frac{d}{4}},H}(t,x))_{t\in\mathbb{R}_+}$ are Gaussian and centered. Moreover, we get from corollary 3.11 that

$$\mathbb{E}\Big(u_{a,H}(ht,x)u_{a,H}(hs,x)\Big) = (2\pi)^{-d/2}\sum_{i=1}^{N} a_i^2 H_i(2H_i-1)$$

$$\times\int_0^{ht}\int_0^{hs}\Big(|u-v|^{2H_i-2}-(u+v)^{2H_i-2}\Big)(ht+hs-(u+v))^{-\frac{d}{2}}dudv.$$

Therefore, changing the variables $u' = \frac{u}{h}$, $v' = \frac{v}{h}$, we immediately get that

$$\mathbb{E}\left(u_{a,H}(ht,x)u_{a,H}(hs,x)\right) = \mathbb{E}\Big(u_{ah^{H-\frac{d}{4}},H}(t,x)u_{ah^{H-\frac{d}{4}},H}(s,x)\Big).$$

This relation immediately implies our statement. □

3.7.5. *Regularity and fractal properties of mild solution*

3.7.5.1. *Regularity of mild solution in time*

In this paragraph, we focus our attention on the behavior of the increments of the mild solution $u_{a,H}(t,x)$ to equation [3.72] with respect to the variable t. Recalling that $H_1 > \frac{1}{2}$ and $d < 4H_1$, as was claimed in equation [3.86], we first state the following technical lemma.

LEMMA 3.16.– *There exist two positive constants $c_1(d,H_1)$ and $c_2(d,H_1)$, depending only on d and H_1, such that for any $s,t\in[0,T]$,*

i)
$$\int_s^t\int_s^t |u-v|^{2H_1-2}(2t-u-v)^{-d/2}dudv = c_1(d,H_1)|t-s|^{2H_1-d/2}.$$

ii)
$$\int_0^s\int_0^s |u-v|^{2H_1-2}\Big[(2t-u-v)^{-d/2}-2(t+s-u-v)^{-d/2}$$
$$+(2s-u-v)^{-d/2}\Big]dvdu \le c_2(d,H_1)|t-s|^{2H_1-d/2}.$$

PROOF.–

i) Changing the variables $u' = u-s, v' = v-s$, and then again changing the variables $x = \dfrac{u'}{t-s}$, $y = \dfrac{v'}{t-s}$, we get

$$\int_s^t \int_s^t |u-v|^{2H_1-2}(2t-u-v)^{-d/2} dudv$$
$$= \int_0^{t-s} \int_0^{t-s} |u'-v'|^{2H_1-2}(2(t-s)-u'-v')^{-d/2} du'dv'$$
$$= |t-s|^{2H_1-d/2} \int_0^1 \int_0^1 |x-y|^{2H_1-2}(2-x-y)^{-d/2} dxdy.$$

Now, changing the variables $x' = x-y, y' = 2-x-y$ and following the same technique as that used in the proof of equation [3.85], we obtain that

$$\int_0^1 \int_0^1 |x-y|^{2H_1-2}(2-x-y)^{-d/2} dxdy < \infty,$$

and deduce the first assertion of the lemma.

ii) Again, changing the variables $u' = s-u, v' = s-v$, and then taking $x = \frac{u'}{t-s}$, $y = \frac{v'}{t-s}$, we get

$$J = \int_0^s \int_0^s |u-v|^{2H_1-2}\Big[(2t-u-v)^{-d/2} - 2(t+s-u-v)^{-d/2}$$
$$+(2s-u-v)^{-d/2}\Big] dvdu$$
$$= \int_0^s \int_0^s |u'-v'|^{2H_1-2}\Big[(2(t-s)+u'+v')^{-d/2}$$
$$-2(t-s+u'+v')^{-d/2} + (u'+v')^{-d/2}\Big] dv'du'$$

$$= |t-s|^{2H_1-d/2} \int_0^{\frac{s}{t-s}} \int_0^{\frac{s}{t-s}} |x-y|^{2H_1-2}\Big[(2+x+y)^{-d/2}$$
$$-2(1+x+y)^{-d/2} + (x+y)^{-d/2}\Big] dxdy$$
$$\le |t-s|^{2H_1-d/2} \int_0^{+\infty} \int_0^{+\infty} |x-y|^{2H_1-2}\Big[(2+x+y)^{-d/2}$$
$$-2(1+x+y)^{-d/2} + (x+y)^{-d/2}\Big] dxdy.$$

The last double integral can be written as the sum of the following two terms:

$$K = \int_0^{+\infty}\int_0^{+\infty} |x-y|^{2H_1-2}\Big[(2+x+y)^{-d/2} -2(1+x+y)^{-d/2} + (x+y)^{-d/2}\Big]dxdy = K_1 + K_2,$$

where

$$K_1 = \int_0^{+\infty}\int_0^{y} |x-y|^{2H_1-2}\Big[(2+x+y)^{-d/2} - 2(1+x+y)^{-d/2} + (x+y)^{-d/2}\Big]dxdy$$

and

$$K_2 = \int_0^{+\infty}\int_y^{\infty} |x-y|^{2H_1-2}\Big[(2+x+y)^{-d/2} - 2(1+x+y)^{-d/2} + (x+y)^{-d/2}\Big]dxdy.$$

Now making the change of variables $x' = x - y, y' = x + y$ in the double integral K_1, we get

$$K_1 = \int_0^{\infty}\int_0^{y'} (x')^{2H_1-2}\Big[(2+y')^{-d/2} - 2(1+y')^{-d/2} + (y')^{-d/2}\Big]dx'dy' = \int_0^{\infty} \frac{1}{2H_1-1}(y')^{2H_1-1}\Big[(2+y')^{-d/2} - 2(1+y')^{-d/2} + (y')^{-d/2}\Big]dy'.$$

The function

$$f : y \mapsto y^{2H_1-1}\Big[(2+y)^{-d/2} - 2(1+y)^{-d/2} + (y)^{-d/2}\Big]$$

is obviously continuous on the interval $(0, +\infty)$. Moreover, on the one hand, when $y \to 0$, because $d < 4H_1$ and $H_1 > 1/2$, we have

$$f(y) \sim (2^{-d/2} - 2)y^{2H_1-1} + y^{2H_1-1-d/2} \sim y^{2H_1-1-d/2}$$

and $\int_0^a y^{2H_1-1-d/2}dy < \infty$ for any $a > 0$.

On the other hand, when $y \to +\infty$,

$$f(y) = y^{2H_1-1-d/2}\Big((1+2/y)^{-d/2} - 2(1+1/y)^{-d/2} + 1\Big)$$
$$\sim y^{2H_1-3-d/2},$$

and $\int_a^\infty y^{2H_1-3-d/2}dy < \infty$ for any $a > 0$ because $2H_1 - 3 - d/2 < -1$. From all this, we get $K_1 < \infty$.

Following the same technique, we show that $K_2 < \infty$ and get the second assertion of the lemma. □

The main result of this is the following proposition.

PROPOSITION 3.10.– *There exist two positive constants C_3 and C_4 such that for any $s, t \in [0, T]$ and any $x \in \mathbb{R}^d$,*

$$C_3|t-s|^{2H_1-\frac{d}{2}} \le \mathbb{E}\left|u_{a,H}(t,x) - u_{a,H}(s,x)\right|^2 \le C_4|t-s|^{2H_1-\frac{d}{2}}.$$

PROOF.– We have

$$\mathbb{E}\left|u_{a,H}(t,x) - u_{a,H}(s,x)\right|^2 = R_{u,a,H}(t,t) - 2R_{u,a,H}(t,s) + R_{u,a,H}(s,s),$$

where $R_{u,a,H}$ denotes the covariance of the process $u_{a,H}$ with respect to the time variable for fixed $x \in \mathbb{R}^d$. Proposition 3.8 implies that

$$R_{u,a,H}(t,s) = \mathbb{E}(u_{a,H}(t,x)u_{a,H}(s,x))$$
$$= (2\pi)^{-d/2}\int_0^t\int_0^s \frac{\partial^2 \mathrm{Cov}(S_u^H, S_v^H)}{\partial u \partial v}(t+s-u-v)^{-d/2}dudv$$

for any $s, t \in [0, T]$. Therefore,

$$\mathbb{E}\left|u_{a,H}(t,x) - u_{a,H}(s,x)\right|^2$$
$$= (2\pi)^{-d/2}\int_0^t\int_0^t \frac{\partial^2 \mathrm{Cov}(S_u^H, S_v^H)}{\partial u \partial v}(2t-u-v)^{-d/2}dudv$$
$$-2(2\pi)^{-d/2}\int_0^t\int_0^s \frac{\partial^2 \mathrm{Cov}(S_u^H, S_v^H)}{\partial u \partial v}(t+s-u-v)^{-d/2}dudv$$
$$+(2\pi)^{-d/2}\int_0^s\int_0^s \frac{\partial^2 \mathrm{Cov}(S_u^H, S_v^H)}{\partial u \partial v}(2s-u-v)^{-d/2}dudv.$$

The last relation can be rewritten as

$$\mathbb{E}\left|u_{a,H}(t,x)-u_{a,H}(s,x)\right|^2 = A_{a,H}(t,s)+B_{a,H}(t,s)+C_{a,H}(t,s),$$

where

$$A_{a,H}(t,s) = (2\pi)^{-d/2}\int_s^t\int_s^t \frac{\partial^2 \mathrm{Cov}(S_u^H, S_v^H)}{\partial u \partial v}(2t-u-v)^{-d/2}dudv,$$

$$B_{a,H}(t,s) = (2\pi)^{-d/2}\int_0^s\int_0^s \frac{\partial^2 \mathrm{Cov}(S_u^H, S_v^H)}{\partial u \partial v}\Big[(2t-u-v)^{-d/2} \\ -2(t+s-u-v)^{-d/2}+(2s-u-v)^{-d/2}\Big]dudv,$$

and

$$C_{a,H}(t,s) = 2(2\pi)^{-d/2}\int_s^t\int_0^s \frac{\partial^2 \mathrm{Cov}(S_u^H, S_v^H)}{\partial u \partial v} \\ \times\Big[(2t-u-v)^{-d/2}-2(t+s-u-v)^{-d/2}\Big]dudv.$$

Because $C_{a,H}(t,s)\le 0$, from inequality [3.71], we have

$$\mathbb{E}\left|u_{a,H}(t,x)-u_{a,H}(s,x)\right|^2 \\ \le c\Big[\int_s^t\int_s^t |u-v|^{2H_1-2}(2t-u-v)^{-d/2}dudv+\int_0^s\int_0^s |u-v|^{2H_1-2} \\ \times\Big((2t-u-v)^{-d/2}-2(t+s-u-v)^{-d/2}+(2s-u-v)^{-d/2}\Big)dudv\Big], \quad [3.87]$$

where c denotes a positive constant. Consequently, the upper bound follows from equations [3.86] and [3.87] and lemma 3.16.

Let us now establish the lower bound. The mild solution has the following form:

$$u_{a,H}(t,x) := \int_0^t\int_{\mathbb{R}^d} G(t-v,x-y)W_{a,H}(dv,dy). \quad [3.88]$$

Therefore, for any $x \in \mathbb{R}^d$ and $(s,t) \in [0,T]^2$,

$$u(t,x) - u(s,x) = \int_0^T \int_{\mathbb{R}^d} \Big(G(t-\sigma, x-y)\mathbf{1}_{(0,t)}(\sigma) - G(s-\sigma, x-y)\mathbf{1}_{(0,s)}(\sigma) \Big) W_{a,H}(d\sigma, dy). \qquad [3.89]$$

From formula [3.75], we get:

$$u_{a,H}(t,x) - u_{a,H}(s,x) = \sum_{i=1}^{N} D_i(a_i, H_i) \int_{\mathbb{R}} \int_{\mathbb{R}^d} F_i(\sigma, y) W_i(d\sigma, dy), \qquad [3.90]$$

where

$$D_i(a_i, H_i) = a_i C(H_i)(H_i - 1/2), \; 1 \leq i \leq N,$$

and

$$F_i(\sigma, y) = \Big(\int_{\mathbb{R}} G(t-u, x-y)\mathbf{1}_{(0,t)}(u) T_i(u,\sigma) du - \int_{\mathbb{R}} G(s-u, x-y)\mathbf{1}_{(0,s)}(u) T_i(u,\sigma) du \Big),$$

with

$$T_i(u,\sigma) = (u-\sigma)_+^{H_i - \frac{3}{2}} + (u+\sigma)_-^{H_i - \frac{3}{2}}.$$

Relation [3.90] and the independence of the Wiener processes W_i allow us to write

$$\begin{aligned} &\mathbb{E}\Big(u_{a,H}(t,x) - u_{a,H}(s,x) \Big)^2 \\ &= \sum_{i=1}^{N} D_i(a_i, H_i)^2 \mathbb{E} \left(\int_{\mathbb{R}} \int_{\mathbb{R}^d} F_i(\sigma, y) W_i(d\sigma, dy) \right)^2 \\ &\geq D_1(a_i, H_1)^2 \mathbb{E} \left(\int_{\mathbb{R}} \int_{\mathbb{R}^d} F_1(\sigma, y) W_1(d\sigma, dy) \right)^2. \end{aligned}$$

It follows from the isometry of the Wiener process W that

$$\begin{aligned} \mathbb{E} \left(\int_{\mathbb{R}} \int_{\mathbb{R}^d} F_1(\sigma, y) W_1(d\sigma, dy) \right)^2 &= \int_{\mathbb{R}} \int_{\mathbb{R}^d} F_1^2(\sigma, y) d\sigma dy \\ &\geq \int_s^t \int_{\mathbb{R}^d} F_1^2(\sigma, y) d\sigma dy. \end{aligned}$$

Because $T_1(\sigma,u) = (u-\sigma)^{H_1-3/2}\mathbf{1}_{u\geq\sigma} + (-u-\sigma)^{H_1-3/2}\mathbf{1}_{u\leq-\sigma}$, we have

$$F_1(\sigma,y) = \int_0^t G(t-u,x-y)T_1(u,\sigma)du - \int_0^s G(s-u,x-y)T_1(u,\sigma)du$$
$$= \int_\sigma^t G(t-u,x-y)(u-\sigma)^{H_1-3/2}du,$$

for any $\sigma \in [s,t]$. Hence,

$$\mathbb{E}\Bigg(\int_{\mathbb{R}}\int_{\mathbb{R}^d} F_1(\sigma,y)W_1(d\sigma,dy)\Bigg)^2$$
$$= \int_s^t\int_{\mathbb{R}^d}\left(\int_\sigma^t G(t-u,x-y)(u-\sigma)^{H_1-3/2}du\right)^2 d\sigma dy.$$

Therefore, for any $0 \leq s \leq t \leq T$, it holds that

$$\mathbb{E}\Bigg(\int_{\mathbb{R}}\int_{\mathbb{R}^d} F_1(\sigma,y)W_1(d\sigma,dy)\Bigg)^2$$
$$= \int_s^t\int_{\mathbb{R}^d}\int_\sigma^t\int_\sigma^t G(t-u,x-y)(u-\sigma)^{H_1-3/2}G(t-v,x-y) \qquad [3.91]$$
$$\times (v-\sigma)^{H_1-3/2}dvdudyd\sigma = \int_s^t\int_s^t\int_{\mathbb{R}^d} G(t-u,x-y)G(t-v,x-y)dy$$
$$\times \int_s^{u\wedge v}(u-\sigma)_+^{H_1-3/2}(v-\sigma)_+^{H_{i_0}-3/2}d\sigma dvdu,$$

where in the last equality, we have used the fact that

$$(s\leq\sigma\leq t,\ \sigma\leq u\leq t,\ \sigma\leq v\leq t) \Longleftrightarrow (s\leq u\leq t,\ s\leq v\leq t,\ s\leq\sigma\leq u\wedge v).$$

Relations [3.91] and [3.84] imply that

$$\mathbb{E}\Bigg(\int_{\mathbb{R}}\int_{\mathbb{R}^d} F_1(\sigma,y)W_1(d\sigma,dy)\Bigg)^2 = \left(\frac{\pi}{2}\right)^{d/2}\int_s^t\int_s^t (2t-u-v)^{-d/2} \qquad [3.92]$$
$$\times \int_s^{u\wedge v}(u-\sigma)_+^{H_1-3/2}(v-\sigma)_+^{H_1-3/2}d\sigma dvdu$$

The change of variable $z = \dfrac{u \wedge v - \sigma}{u \vee v - \sigma}$ and an easy calculation allow us to get

$$\begin{aligned} &\int_s^{u\wedge v} (u-\sigma)^{H_1-3/2}(v-\sigma)^{H_1-3/2} d\sigma \\ &= |u-v|^{2H_1-2} \int_0^{\frac{u\wedge v - s}{u \vee v - s}} (1-z)^{1-2H_1} z^{H_1-3/2} dz. \end{aligned} \tag{3.93}$$

Then, it follows from equations [3.92] and [3.93] that

$$\begin{aligned} &\mathbb{E}\left(\int_{\mathbb{R}} \int_{\mathbb{R}^d} F_1(\sigma, y) W_1(d\sigma, dy) \right)^2 \\ &= \left(\frac{\pi}{2}\right)^{d/2} \int_s^t \int_s^t (2t-u-v)^{-d/2} |u-v|^{2H_1-2} \\ &\int_0^{\frac{u\wedge v - s}{u\vee v - s}} (1-z)^{1-2H_1} z^{H_1-3/2} dz dv du \end{aligned}$$

Now, changing the variables $u - s = u'$ and $v - s = v'$, we obtain:

$$\begin{aligned} &\mathbb{E}\left(\int_{\mathbb{R}} \int_{\mathbb{R}^d} F_1(\sigma, y) W_1(d\sigma, dy) \right)^2 \\ &= \left(\frac{\pi}{2}\right)^{d/2} \int_0^{t-s} \int_0^{t-s} (2(t-s)-u-v)^{-d/2} \\ &\times |u-v|^{2H_1-2} \int_0^{\frac{u\wedge v}{u \vee v}} (1-z)^{1-2H_1} z^{H_1-3/2} dz dv du \end{aligned}$$

Finally, changing the variables $\tilde{u} = \frac{u}{t-s}$ and $\tilde{v} = \frac{v}{t-s}$, we get that

$$\mathbb{E}\left(\int_{\mathbb{R}} \int_{\mathbb{R}^d} F_1(\sigma, y) W_1(d\sigma, dy) \right)^2 = D(d, H_1)(t-s)^{2H_1 - \frac{d}{2}},$$

where $D(d, H_1)$ is the constant defined by

$$\begin{aligned} D(d, H_1) = c \int_0^1 du \int_0^1 dv (2-u-v)^{-d/2} |u-v|^{2H_1-2} \\ \cdot \int_0^{\frac{u\wedge v}{u\vee v}} (1-z)^{1-2H_1} z^{H_1-3/2} dz. \end{aligned}$$

The constant $D(d, H_1)$ is clearly finite because $H_1 > \frac{1}{2}$. □

REMARK 3.10.– Proposition 3.10 tells us that the process $(u_{a,H}(\cdot, x))$ is in a quasi-helix (see definition 1.11 in Chapter 1) of index $I = H_1 - \dfrac{d}{4}$.

In particular, as an immediate consequence of proposition 3.10, we get the following statement.

COROLLARY 3.12.– *For any $x \in \mathbb{R}^d$, the process $t \to u_{a,H}(t,x)$ is Hölder continuous of any order $\delta \in (0, H_1 - \frac{d}{4})$.*

As a second consequence of proposition 3.10, by proceeding as in the proof of theorem 3.5, we get the following result.

COROLLARY 3.13.– *For any $x \in \mathbb{R}^d$ and any t_0,*

$$\lim_{\epsilon \to 0} \sup_{t \in [t_0 - \epsilon, t_0 + \epsilon]} \left| \frac{u_{a,H}(t,x) - u_{a,H}(t_0,x)}{t - t_0} \right| = +\infty$$

a.s. Consequently, the trajectories of the process $u_{a,H}(\cdot, x)$ are not differentiable.

3.7.5.2. *Fractal characteristics of the sample paths*

For fixed $x \in \mathbb{R}^d$, we denote the range of the restriction of the process $u_{a,H}(\cdot, x)$ on $[0,T]$ by

$$u_{a,H}([0,T], x) = \{u_{a,H}(t,x); t \in [0,T]\}, \tag{3.94}$$

and its graph by

$$\mathrm{Grf}_T u_{a,H}(\cdot, x) = \{(t, u_{a,H}(t,x)); t \in [0,T]\}, \tag{3.95}$$

where $T > \epsilon > 0$. The aim of this paragraph is to study the Hausdorff dimension of the sets defined just above (see definition A.17 in the Appendix).

Let us start our study with the set $\mathrm{Grf}_T u_{a,H}(\cdot, x)$. Throughout the subsequent part of the section, C denotes a generic positive constant that may be different from one line to another.

LEMMA 3.17.– *For any $T > 0$, with probability 1,*

$$\dim \mathrm{Grf}_T u_{a,H}(\cdot, x) = 2 - H_1 + \frac{d}{4},$$

where dim *denotes the Hausdorff dimension.*

PROOF.– It follows from corollary 3.12 that for any $T > 0$ and $x \in \mathbb{R}^d$, $u_{a,H}(\cdot, x)$ has a modification, whose sample paths are Hölder continuous of order $\gamma < H_1 - \frac{d}{4}$ on the interval $[0, T]$. Therefore, it follows from lemma A.10 in the Appendix that for any $T > 0$, with probability 1,

$$\dim \mathrm{Grf}_T u_{a,H}(\cdot, x) \leq 2 - H_1 + \frac{d}{4}.$$

Now, in order to get the lower bound, following the same technique used in the proof of theorem 3.5, we only need to show that for any $T > 0$

$$\int_0^T \int_0^T \mathbb{E}\left((|s-t| + |u_{a,H}(s,x) - u_{a,H}(t,x)|)^{-\gamma}\right) dsdt < +\infty. \qquad [3.96]$$

Because $u_{a,H}(\cdot, x)$ is a centered Gaussian process, we easily check that for all $(s,t) \in \mathbb{R}^2$, $s \neq t$ and for any real $\gamma > 1$, it holds that

$$\mathbb{E}\left((|s-t| + |u_{a,H}(t,x) - u_{a,H}(s,x)|)^{-\gamma}\right) \leq C|t-s|^{1-\gamma}\sigma_{a,H,x}^{-1}(s,t), [3.97]$$

where

$$\sigma_{a,H,x}^2(s,t) = \mathbb{E}\Big(u_{a,H}(t,x) - u_{a,H}(s,x)\Big)^2.$$

Now, using equation [3.97] and proposition 3.10, we get

$$\int_0^T \int_0^T \mathbb{E}\Big(|s-t| + |u_{a,H}(s,x) - u_{a,H}(t,x)|^{-\gamma}\Big) dsdt$$
$$\leq C \int_0^T \int_0^T |t-s|^{1-\gamma}\sigma_{a,H,x}^{-1}(s,t) dsdt \leq C \int_0^T \int_0^T |t-s|^{1+\frac{d}{4}-H_1-\gamma} dsdt.$$

Because $\gamma \in (1, 2 - H_1 + \frac{d}{4})$, the last double integral is finite, and the proof follows. □

In the following lemma, we will give the Hausdorff dimension of the set $u_{a,H}([0,T], x)$.

LEMMA 3.18.– *For any $T > 0$, with probability 1,*

$$\dim u_{a,H}([0,T], x) = 1.$$

PROOF.– It follows from corollary 3.12 that for any $x \in \mathbb{R}^d$, we have

$$\mathbb{P}\left(\omega \in \Omega | t \longmapsto u_{a,H}(t,x) \text{ is continuous on } [0,T]\right) = 1.$$

Thus,

$$\mathbb{P}\left(\omega \in \Omega | u_{a,H}([0,T],x) = [a_\omega, b_\omega];\ (a_\omega, b_\omega) \in \mathbb{R}^2; a_\omega < b_\omega\right) = 1.$$

From corollary A.1 in the Appendix, we get $\dim[a,b] = 1$, for any real $a < b$. Consequently,

$$\mathbb{P}\left(\omega \in \Omega | \dim u_{a,H}([0,T],x) = 1\right) = 1. \qquad \square$$

3.8. Exercises

EXERCISE 3.1.– *Let us consider $a = (a_1, \ldots, a_N) \in \mathbb{R}^N$ with non-zero coordinates and $H = (H_1, \ldots, H_N) \in (0,1)^N$ such that $H_j \neq 1/2$ exists. Prove that $(M_t^H(a))_{t \in \mathbb{R}_+}$ is not a Markov process.*

Hint: Apply the criterion for Gaussian process to be a Markov process presented in [3.6] and compare

$$\mathrm{Cov}\left(M_{1/2}^H, M_{3/2}^H\right)\mathrm{Cov}\left(M_1^H, M_1^H\right)$$

and

$$\mathrm{Cov}\left(M_{1/2}^H, M_1^H\right)\mathrm{Cov}\left(M_1^H, M_{3/2}^H\right).$$

EXERCISE 3.2.– *Let us consider $T > 0$ and $\gamma < H_1 = \min\{H_i; i \in \{1, \ldots, N\}\}$. Prove that the mixed fractional Brownian motion has a modification, the trajectories of which have a Hölder continuity property of order γ on the interval $[0,T]$.*

EXERCISE 3.3.– *Show that for any $H \in (0,1)^N$ the trajectories of a mixed fractional Brownian motion with probability 1 are not differentiable.*

EXERCISE 3.4.– *Denoting X^n and X^H the processes defined, respectively, by relations [3.41] and [3.42], establish that*

$$\limsup_{n \to \infty} n^{2H \wedge 1} \mathbb{E} \|X^n - X^H\|^2_{L^2([0,1])} \leq c,$$

where c denotes a positive constant.

EXERCISE 3.5.– *Consider a stochastic process $X = \{X_t, t \geq 0\}$ defined on a probability space $(\Omega, \mathcal{F}, \mathbb{P})$, and $\mathcal{F}^X$ its natural filtration (see, for instance, example A.7 in the Appendix). Prove that $\mathcal{F}^X$ is the smallest filtration with respect to which X is adapted.*

EXERCISE 3.6.– *Consider a measurable space* $(\Omega, \mathcal{F})$ *equipped with a filtration* $\mathbb{F} = \{\mathcal{F}_t, t \geq 0\}$ *and a stopping time* T *of the filtration* $\mathbb{F}$. *Prove that* $\{T < t\} \in \mathcal{F}_t$, *for any* $t \geq 0$.

EXERCISE 3.7.– *Consider the deterministic process*

$$X_t = \begin{cases} 0 \text{ if } t \in [0, T/2] \\ 1 \text{ if } t \in (T/2, T] \end{cases}$$

Prove that X *is a weak semi-martingale and that* X *is not a semi-martingale.*

Appendix

A.1. Elements of calculus. Some functional classes and properties of functions

A.1.1. *First- and higher-order variations of functions*

A.1.1.1. *Variation of the first order*

Fix $t > 0$ and consider a partition $\pi_m = \{t_0, t_1, \ldots, t_m\}$ consisting of the points $0 = t_0 < t_1 < \cdots < t_m = t$. The mesh of the partition π_m is defined by

$$\|\pi_m\| = \max_{1 \le k \le m} |t_k - t_{k-1}|.$$

Another partition π_n of $[0, t]$ is called a refinement of π_m, if any point of π_m is a point of π_n.

Let $f = \{f_t, t \ge 0\}$ be a real-valued measurable function. Create the sum

$$V_t(\pi_m, f) = \sum_{k=1}^{m} |f_{t_k} - f_{t_{k-1}}|.$$

REMARK A.1.– If π_n is a *refinement* of π_m, then $V_t(\pi_n, f) \ge V_t(\pi_m, f)$.

DEFINITION A.1.– *We say that the function f is of finite variation on the interval $[0, t]$ if*

$$V_t = V_t(f) := \sup_{\pi_m} V_t(\pi_m, f) < \infty.$$

The value $V_t(f)$ is called a variation of f on $[0, t]$.

REMARK A.2.–

i) Assume that $V_T(f) < \infty$ for some $T > 0$. Because V_t is a non-decreasing function of t, then it is finite for any $t \in [0,T]$.

ii) Regardless of the value of $V_T(f)$, there exists a sequence of refining partitions $\pi_{m_n} = \{t_0^n, t_1^n, \dots, t_m^n\}$ such that $\|\pi_m\| \to 0$ and

$$\lim_{n\to\infty} V_t(\pi_{m_n}, f) = V_t(f).$$

This follows from the definition of supremum and remark A.1. The situation with the variations of the higher order is different. The difference between these quantities becomes clear after considering lemma A.1.

A.1.1.2. *Variations of the higher orders*

Let $f = \{f_t, t \geq 0\}$ be a real-valued measurable function, and let $p > 1$. Create the sum

$$V_t^p(\pi_m, f) = \sum_{k=1}^{m} |f_{t_k} - f_{t_{k-1}}|^p.$$

Starting from this definition, we can go in two ways and give two definitions of higher-order variations.

DEFINITION A.2.–

i) Let, for function f and any sequence of partitions $\pi_{m_n} = \{t_0^n, t_1^n, \dots, t_m^n\}$ such that $\|\pi_{m_n}\| \to 0$, there exist a limit $\lim_{n\to\infty} V_t^p(\pi_{m_n}, f)$, which does not depend on the sequence of partitions. Then, we say that function f has on the interval $[0,T]$ the finite p-variation.

ii) Let it hold that, for function f, $\sup_{\pi_m} V_t^p(\pi_m, f) < \infty$. Then, we say that function f has on the interval $[0,T]$ the finite total p-variation.

Let $p = 2$. Then the 2–variation on the interval $[0,t]$ is called the quadratic variation and is denoted by $[f]_t$. The next result demonstrates the difference between p-variation and total p-variation. Note that obviously total p-variation is positive for any non-constant function.

LEMMA A.1.– *If $f : [0,T] \to \mathbb{R}$ is a continuous function of finite variation, then its p-variation on $[0,T]$ equals zero for any $p > 1$.*

PROOF.– Consider a partition $\pi_m = \{t_0, t_1, \dots, t_m\}$ of $[0,T]$. We have that

$$V_T^{(p)}(\pi_m, f) = \sum_{k=1}^{m} |f_{t_k} - f_{t_{k-1}}|^p \leq \sup_{1\leq k\leq m} |f_{t_k} - f_{t_{k-1}}|^{p-1} V_T(f),$$

which tends to zero when $\|\pi_m\| \to 0$ because of the continuity of f and the finiteness of $V_T(f)$. □

A.2. Convex, concave functions and some related questions

DEFINITION A.3.– *Consider a non-empty interval $I \subset \mathbb{R}$. A function $f : I \to \mathbb{R}$ is called convex (respectively concave) if*

$$f((1-\lambda)x + \lambda y) \leq (1-\lambda)f(x) + \lambda f(y),$$

(respectively, $f((1-\lambda)x + \lambda y) \geq (1-\lambda)f(x) + \lambda f(y)$) for any $x, y \in I$ and $\lambda \in (0,1)$.

LEMMA A.2.– *Let f be a real-valued function defined on an interval I. Then, the following assertions are equivalent:*

1) f is concave on I.

2) $\begin{vmatrix} 1 & x & f(x) \\ 1 & y & f(y) \\ 1 & z & f(z) \end{vmatrix} \leq 0$ for all $x < y < z \in I$.

PROOF.– We have that $\begin{vmatrix} 1 & x & f(x) \\ 1 & y & f(y) \\ 1 & z & f(z) \end{vmatrix} \leq 0$ if and only if

$$(z-y)f(x) - (z-x)f(y) + (y-x)f(z) \leq 0$$

for any $x < y < z \in I$. Further, for any $y \in (x,z)$, there exists $\lambda \in (0,1)$ such that $y = (1-\lambda)x + \lambda z$. Therefore, $(z-y)f(x) - (z-x)f(y) + (y-x)f(z) \leq 0$ for any $x < y < z \in I$ if and only if

$$(1-\lambda)f(x) + \lambda f(z) \leq f((1-\lambda)x + \lambda z)$$

for any $x < z \in I$ and $\lambda \in (0,1)$. In turn, this is equivalent to the concavity of f. □

COROLLARY A.1.– *Let f be a real-valued function that is concave on an interval I. Then, for any $x < u \leq v < y \in I$, we have*

$$\frac{f(u)-f(x)}{u-x} \geq \frac{f(v)-f(x)}{v-x} \geq \frac{f(v)-f(y)}{v-y}.$$

PROOF.– Indeed,

$$\frac{f(u)-f(x)}{u-x} \geq \frac{f(v)-f(x)}{v-x} \iff (v-u)f(x)-(v-x)f(u)+(u-x)f(v) \leq 0$$

$$\iff \begin{vmatrix} 1 & x & f(x) \\ 1 & u & f(u) \\ 1 & v & f(v) \end{vmatrix} \leq 0.$$

Therefore, the first inequality in the corollary is a straightforward consequence of lemma A.2.

For the second inequality, we mention that

$$\frac{f(v)-f(x)}{v-x} \geq \frac{f(v)-f(y)}{v-y} \iff (y-v)f(x)-(y-x)f(v)+(v-x)f(y) \leq 0$$

$$\iff \begin{vmatrix} 1 & v & f(v) \\ 1 & x & f(x) \\ 1 & y & f(y) \end{vmatrix} \leq 0.$$

This with lemma A.2 implies the second inequality. □

LEMMA A.3.– *Let for any fixed $s > 0$ and $H \in (0,1)$ the function f be defined as*

$$f(x) = -2^{2H-1}((x+s)^{2H} + s^{2H}) + (x+2s)^{2H} - \left(1-2^{2H-1}\right)x^{2H}, x \geq 0.$$

Then,

i) If $H < \frac{1}{2}$, then the function f is decreasing and takes non-positive values.

ii) If $H > \frac{1}{2}$, then the function f is increasing and takes non-negative values.

PROOF.– Obviously, $f(0) = 0$. Further, its derivative equals

$$f'(x) = 2H\left(-2^{2H-1}(x+s)^{2H-1} + (x+2s)^{2H-1} - (1-2^{2H-1})x^{2H-1}\right).$$

Let $H < \frac{1}{2}$. Then,

$$f'(x) = 2H\Big([(x+2s)^{2H-1} - (x+s)^{2H-1}] + [(1-2^{2H-1})((x+s)^{2H-1} - x^{2H-1})]\Big) \leq 0$$

because both values in the square brackets are non-positive. Therefore, f is decreasing and non-positive.

Now, let $H > \frac{1}{2}$. Then, $f'(0) = 0$. For any fixed $x > 0$, consider

$$\frac{\partial(-2^{2H-1}(x+s)^{2H-1} + (x+2s)^{2H-1} - (1-2^{2H-1})x^{2H-1})}{\partial s}$$
$$= (2H-1)(-2^{2H-1}(x+s)^{2H-2} + 2(x+2s)^{2H-2})$$
$$= (2H-1)(-2^{2H-1}(x+s)^{2H-2} + 2^{2H-1}\left(\frac{x}{2}+s)^{2H-2}\right) \geq 0.$$

It means that $f'(x) > 0$ for any $x > 0$ and consequently f is increasing and non-negative. □

For $H \in [1/2, 1)$ denote by

$$f_1(k) = (j-k+2)^{2H} - 2(j-k+1)^{2H} + (j-k)^{2H}$$

and

$$f_2(k) = (j+k+1)^{2H} - 2(j+k)^{2H} + (j+k-1)^{2H}.$$

LEMMA A.4.– *For any* $1 \leq k \leq j$

i) $f_1(k) \geq 0$ *and* $f_2(k) \geq 0$.

ii) $f_1(k) - f_2(k) > 0$

iii) $f_1(k) - f_2(k) \geq 2H(2H-1)\left((j-k+1)^{2H-2} - (j+k+1)^{2H-2}\right) \geq 0.$

PROOF.–

i) These inequalities are due to the fact that the function $x \longmapsto x^{2H}$ is convex on the interval $[0, +\infty)$.

ii) Let us consider the function g_l defined by

$$g(x) = (x+1)^{2H} - 2\,x^{2H} + (x-1)^{2H}, \quad x \geq 1.$$

Because the function $x \longmapsto x^{2H-1}$ is concave on $[1, +\infty[$, g decreases on this interval and consequently

$$f_1(k) = g(j-k+1) > g(j+k) = f_2(k).$$

iii) For any $a \geq 1$, let us consider the function g_a defined by

$$g_a(x) = (a+x)^{2H} - 2\,a^{2H} + (a-x)^{2H}, \quad 0 \leq x \leq 1 \leq a.$$

On the one hand, we have that $g_a(0) = 0$ and

$$g_a^{'}(x) = 2H\left((a+x)^{2H-1} - (a-x)^{2H-1}\right).$$

Therefore, $g_a'(0) = 0$. On the other hand, it follows from the Taylor–Lagrange theorem that there exists $c \in (0,1)$ such that

$$g_a(1) = g_a(0) + g_a'(0) + 1/2 g_a''(c) = 1/2 g_a''(c).$$

Next, it is easy to check that the function g_a'' increases and, consequently,

$$g_a''(0) \leq g_a''(c) \leq g_a''(1).$$

So, we have

$$2H(2H-1)a^{2H-2} \leq g_a(1) \leq H(2H-1)\Big((a+1)^{2H-2} + (a-1)^{2H-2}\Big),$$

which yields that

$$\begin{aligned} f_1(k) &= g_{j-k+1}(1) \geq 2H(2H-1)\,(j-k+1)^{2H-2} \text{ and} \\ f_2(k) &= g_{j+k}(1) \leq H(2H-1)\Big((j+k+1)^{2H-2} + (j+k-1)^{2H-2}\Big) \\ &\leq 2H(2H-1)(j+k-1)^{2H-2}, \end{aligned}$$

that ends the proof of the lemma. □

A.3. Elements of linear, metric, normed and Hilbert spaces

A.3.1. *Linear spaces*

Introduce the notion of the real linear space or, in other words, of the linear space over the field $\mathbb{R}$ of the real numbers.

DEFINITION A.4.– *Let E be a non-empty set whose elements will be interpreted as the vectors. Introduce the following operations:*

i) Vector addition: $+ : E \times E \to E; \forall (x,y) \in E \times E, x+y \in E$.

ii) Scalar multiplication: $\times : \mathbb{R} \times E \to E; \forall (\lambda, y) \in \mathbb{R} \times E, \lambda y := \lambda \times y \in E$.

We say that $(E, +, \times)$ (or simply E) is a real linear space if the following eight axioms are satisfied:

i) $x+(y+z) = (x+y)+z$, *for any* $x, y, z \in E$ *(associativity of vector addition).*

ii) $x + y = y + x$, *for any* $x, y \in E$ *(commutativity of vector addition).*

iii) There exists an element $0 \in E$ *such that* $x + 0 = 0 + x = x$, *for any* $x \in E$ *(existence of zero element).*

iv) For any $x \in E$, *there exists* $y \in E$ *such that* $x + y = y + x = 0$ *(existence of opposite element).*

v) $\lambda \times (\beta \times x) = (\lambda \times \beta) \times x$, *for any* $\lambda, \beta \in \mathbb{R}$ *and* $x \in \mathbb{R}$ *(associativity of scalar multiplication).*

vi) $\lambda \times (x + y) = \lambda \times x + \lambda \times y$, *for any* $\lambda \in \mathbb{R}$ *and* $x, y \in \mathbb{R}$ *(combined associativity).*

vii) $(\lambda + \beta) \times x = \lambda \times x + \beta \times x$, *for any* $\lambda, \beta \in \mathbb{R}$ *and* $x \in \mathbb{R}$ *(combined associativity).*

viii) $1 \times x = x$, *for any* $x \in \mathbb{E}$ *(existence of scalar unit).*

EXAMPLE A.1.– *Consider a non-empty subset* $\mathbb{T} \subset \mathbb{R}$, *and let*

$$E = \mathcal{C}(\mathbb{T}) = \{f : \mathbb{T} \to \mathbb{R} |\ f \text{ is a continuous function}\},$$

supplied with the operations

$$+ : E \times E \to E; \text{ and } \times : \mathbb{R} \times E \to E,$$

defined, respectively, by

$$(f, g) \longmapsto f + g; \quad (f + g)(t) = f(t) + g(t); \quad \forall t \in \mathbb{T},$$

and

$$(\lambda, f) \longmapsto \lambda \times f; \quad (\lambda \times f)(t) = \lambda f(t)\ \forall t \in \mathbb{T}.$$

Then, E *is a real linear space.*

DEFINITION A.5.– *Let* $(E, +, \times)$ *be a real linear space and* F *be a subset of* E. *We say that* $(F, +, \times)$ *(or simply* F*) is a linear subspace of* E, *if*

i) zero vector 0 *is in* F.

ii) If u *and* v *are elements of* F, *then* $u + v$ *is an element of* F.

iii) If u *is an element of* F *and* λ *is a real number, then* $\lambda \cdot u$ *is an element of* F.

A.3.2. *Normed and semi-normed spaces*

DEFINITION A.6.– *Consider a real linear space E and a function $N : E \to \mathbb{R}$. The function N is called a norm on E if, for all $v, w \in E$ and all $\lambda \in \mathbb{R}$ it holds that*

i) $N(v) \geq 0$,

ii) $N(\lambda v) = |\lambda| N(v)$,

iii) $N(v + w) \leq N(v) + N(w)$

and

iv) $N(v) = 0 \Longleftrightarrow v = 0$.

If N is a norm on E, then the pair (E, N) (or simply E) is called a normed space. If N satisfies only assertions(i), (ii) and (iii), we say that the function N is a semi-norm on E, and the pair (E, N) is called a semi-normed space.

REMARK A.3.– If N is a semi-norm, it can happen that $N(v) = 0$ for some non-zero v.

EXAMPLE A.2.– *Consider the linear space $E = \mathcal{C}(\mathbb{R})$ of real-valued functions, continuous on $\mathbb{R}$. Let $N : f \in E \longmapsto |f(0)|$. Then, N is a semi-norm, which is not a norm.*

REMARK A.4.– We note that a semi-norm N is a norm if $N(v) = 0$ is equivalent to $v = 0$.

EXAMPLE A.3.– *Let $T > 0$. Then, $N : x \longmapsto \sup_{s \leq T} |x(s)|$ is a norm on $C([0, T])$. Indeed, for any $x, y \in C([0, T])$ and $\lambda \in \mathbb{R}$, we have that*

i) $N(x) \geq 0$.

ii)
$$N(\lambda x) = \sup_{s \leq T} |\lambda x(s)| = |\lambda| \sup_{s \leq T} |x(s)| = |\lambda| N(x),$$

iii)
$$\begin{aligned} N(x + y) &= \sup_{s \leq T} |(x + y)(s)| \leq \sup_{s \leq T} (|x(s)| + |y(s)|) \\ &\leq \sup_{s \leq T} |x(s)| + \sup_{s \leq T} |y(s)| = N(x) + N(y) \end{aligned}$$

and

iv)
$$N(x) = 0 \Longleftrightarrow \sup_{s \leq T} |x(s)| = 0 \Longleftrightarrow x(s) = 0, \forall s \leq T \Longleftrightarrow x = 0.$$

DEFINITION A.7.– *A measurable linear space is a pair $(E, \mathcal{B})$ of a real linear space E and a σ-algebra $\mathcal{B}$ on E such that the functions $(x, y) \longmapsto x + y$ and $(\lambda, x) \longmapsto \lambda x$*

are, respectively, measurable from $(E \times E, \mathcal{B} \otimes \mathcal{B})$ *to* $(E, \mathcal{B})$ *and from* $(\mathbb{R} \times E, \mathcal{B}(\mathbb{R}) \otimes \mathcal{B})$ *to* $(E, \mathcal{B})$, *where* $\mathcal{B}(\mathbb{R})$ *denotes the Borel* σ*-algebra on* $\mathbb{R}$.

EXAMPLE A.4.– *Let* $\mathbb{T}$ *be a non-empty subset of* $\mathbb{R}$ *and consider the space of continuous functions* $E = C(\mathbb{T})$, *together with the Borel* σ*-algebra* $\mathcal{B}$ *on* E. *Then,* $(E, \mathcal{B})$ *is a measurable linear space.*

DEFINITION A.8.– *Consider a measurable space* $(E, \mathcal{B})$ *and a function* $N : (E, \mathcal{B}) \to (\overline{\mathbb{R}}, \mathcal{B}(\overline{\mathbb{R}}))$. *Function* N *is called a pseudo-semi-norm on* E *if* $N^{-1}(\mathbb{R})$ *is a linear subspace of* E *such that the restriction of* N *on* $N^{-1}(\mathbb{R})$ *is a semi-norm.*

EXAMPLE A.5.– *For a fixed* $T > 0$, *consider*

$$x \longmapsto v(x) = \left(\sup_{n \geq 1} \sum_{j=1}^{n} \left[x\left(\frac{jT}{n}\right) - x\left(\frac{(j-1)T}{n}\right) \right]^2 \right)^{1/2}, \quad x \in C([0,T]).$$

Then, v *is a pseudo-semi-norm on* $C([0,T])$. Indeed, consider $x, y \in C([0,T])$ and $\lambda \in \mathbb{R}$. It is obvious that $v(x) \geq 0$ and $v(\lambda x) = |\lambda| v(x)|$. Now,

$$\begin{aligned}
v^2(x+y) &= \sup_{n \geq 1} \sum_{j=1}^{n} \left[(x+y)\left(\frac{jT}{n}\right) - (x+y)\left(\frac{(j-1)T}{n}\right) \right]^2 \\
&= \sup_{n \geq 1} \sum_{j=1}^{n} \left[x\left(\frac{jt}{n}\right) - x\left(\frac{(j-1)t}{n}\right) + y\left(\frac{jt}{n}\right) - y\left(\frac{(j-1)t}{n}\right) \right]^2 \\
&= \sup_{n \geq 1} \left(\sum_{j=1}^{n} \left[x\left(\frac{jt}{n}\right) - x\left(\frac{(j-1)t}{n}\right) \right]^2 + \sum_{j=1}^{n} \left[y\left(\frac{jt}{n}\right) - y\left(\frac{(j-1)t}{n}\right) \right]^2 \right. \\
&\quad \left. + 2 \sum_{j=1}^{n} \left[x\left(\frac{jt}{n}\right) - x\left(\frac{(j-1)t}{n}\right) \right] \left[y\left(\frac{jt}{n}\right) - y\left(\frac{(j-1)t}{n}\right) \right] \right) \\
&\leq \sup_{n \geq 1} \left(\sum_{j=1}^{n} \left[x\left(\frac{jt}{n}\right) - x\left(\frac{(j-1)t}{n}\right) \right]^2 + \sum_{j=1}^{n} \left[y\left(\frac{jt}{n}\right) - y\left(\frac{(j-1)t}{n}\right) \right]^2 \right. \\
&\quad + 2 \left(\sum_{j=1}^{n} \left[x\left(\frac{jt}{n}\right) - x\left(\frac{(j-1)t}{n}\right) \right]^2 \right)^{1/2}
\end{aligned} \qquad \text{[A.1]}$$

$$\times \left(\sum_{j=1}^{n}\left[y\left(\frac{jt}{n}\right)-y\left(\frac{(j-1)t}{n}\right)\right]^2\right)^{1/2}\Bigg)$$

$$= \sup_{n\geq 1}\left[\left(\sum_{j=1}^{n}\left[x\left(\frac{jt}{n}\right)-x\left(\frac{(j-1)t}{n}\right)\right]^2\right)^{1/2}\right.$$

$$\left.+\left(\sum_{j=1}^{n}\left[y\left(\frac{jt}{n}\right)-y\left(\frac{(j-1)t}{n}\right)\right]^2\right)^{1/2}\right]^2$$

$$\leq \left[\sup_{n\geq 1}\left(\sum_{j=1}^{n}\left[x\left(\frac{jt}{n}\right)-x\left(\frac{(j-1)t}{n}\right)\right]^2\right)^{1/2}\right.$$

$$\left.+\sup_{n\geq 1}\left(\sum_{j=1}^{n}\left[y\left(\frac{jt}{n}\right)-y\left(\frac{(j-1)t}{n}\right)\right]^2\right)^{1/2}\right]^2$$

$$= (v(x)+v(y))^2.$$

Recalling that $v(x) \geq 0$ for any $x \in C([0,T])$, we deduce from [A.1] that $v(x + y) \leq v(x) + v(y)$. This, with the fact that the zero function is in $C([0,T])$, imply that the set $v^{-1}(\mathbb{R}) = v^{-1}(\mathbb{R}_+)$ is a linear subspace of $C([0,T])$. From all this, we deduce that v is a pseudo-semi-norm on $C([0,T])$.

EXAMPLE A.6.– *Consider $t > 0$. Using the same technique that was used in example A.5, one can show that $N : f \longmapsto V_t(f)$ and $N : f \longmapsto [f]_t$ are both pseudo-semi-norms on $C([0,t])$.*

A.3.3. *Inner product spaces*

Consider a linear space E over $\mathbb{R}$.

DEFINITION A.9.– *A function $\langle \cdot,\cdot \rangle \to \mathbb{R}$ is called an inner product if for any $x, y, z \in E$ and $\alpha, \beta \in \mathbb{R}$,*

i) $\langle x, y\rangle = \langle y, x\rangle$.

ii) $\langle x, \alpha y + \beta z\rangle = \alpha\langle x, y\rangle + \beta\langle x, z\rangle$.

iii) $\langle x, x\rangle \geq 0$.

iv) $\langle x, x\rangle = 0 \Longleftrightarrow x = 0$.

The space E with the inner product $\langle \cdot,\cdot \rangle$ is called a real inner product space.

LEMMA A.5.– *(Schwarz inequality). Consider a real inner product space* $(E, \langle \cdot, \cdot \rangle)$. *If we define* $\|x\| = \sqrt{\langle x, x \rangle}$ *for any* $x \in E$, *then*

$$|\langle x, y \rangle \leq \|x\| \cdot \|y\|.$$

LEMMA A.6.– *Consider* $(E, \langle \cdot, \cdot \rangle)$ *a real inner product space. The function* $\| \cdot \|$ *defined by*

$$\| \cdot \| : E \to \mathbb{R}_+ : x \longmapsto \|x\| = \sqrt{\langle x, x \rangle}$$

is a norm. It is the norm induced by the inner product $\langle \cdot, \cdot \rangle$.

PROOF.– For any $\alpha \in \mathbb{R}$, $x, y \in E$, we get from the definition of an inner product and from the Schwarz inequality that

$$\|x\| = 0 \Longleftrightarrow \langle x, x \rangle = 0 \Longleftrightarrow x = 0,$$

$$\|\alpha x\| = \sqrt{\langle \alpha x, \alpha x \rangle} = \sqrt{\alpha^2 \langle x, x \rangle} = |\alpha| \|x\|,$$

and

$$\|x + y\|^2 = \langle x + y, x + y \rangle = \|x\|^2 + 2\langle x, y \rangle + \|y\|^2$$

$$\leq \|x\|^2 + 2\|x\| \|y\| + \|y\|^2 = (\|x\| + \|y\|)^2. \qquad \square$$

A.3.4. *Metric and semi-metric spaces*

DEFINITION A.10.– *Consider a non-empty set* E *and a function* $d : E \times E \to \mathbb{R}$. *The function* d *is called metric if for any* $x, y, z \in \mathbb{R}$, *the following assertions hold:*

i) $d(x, y) \geq 0$ *and* $d(x, y) = 0 \Longleftrightarrow x = y$.

ii) $d(x, y) = d(y, x)$.

iii) $d(x, y) \leq d(x, z) + d(z, y)$.

If d *is a metric on* E, *the pair* (E, d) *is called a metric space.*

REMARK A.5.– If the function $d : E \times E \to \mathbb{R}$ satisfies only the assertions

$$d(x, y) \geq 0, \; d(x, y) = d(y, x) \text{ and } d(x, y) \leq d(x, z) + d(z, y) \text{ for any } x, y, z \in \mathbb{R},$$

the function d is called semi-metric, and the pair (E, d) is called a semi-metric space.

LEMMA A.7.– *Consider a real inner product space $(E, \langle\cdot,\cdot\rangle)$ and the induced norm $\|\cdot\|$. The function $d : E \times E \to \mathbb{R}_+$ defined by the equality*

$$d(x, y) = \|x - y\| = \sqrt{\langle x - y, x - y\rangle},$$

is a metric. It is the metric induced by the inner product $\langle\cdot,\cdot\rangle$. Thus, (E, d) is a metric space.

PROOF.– For any $x, y, z \in E$, we get from the properties of the inner product and from the definition of the function d that

$$d(x, y) = \sqrt{\langle x - y, x - y\rangle} \geq 0,$$

$$d(x, y) = 0 \Longleftrightarrow \langle x - y, x - y\rangle = 0 \Longleftrightarrow x - y = 0 \Longleftrightarrow x = y,$$

$$d(x, y) = \sqrt{\langle x - y, x - y\rangle} = \sqrt{\langle y - x, y - x\rangle} = d(y, x),$$

and

$$d(x, y) = \|x - y\| \leq \|x - z\| + \|z - y\| = d(x, z) + d(z, y). \qquad \square$$

Applying the same technique as in the proof of lemma A.7, we also get the following result.

LEMMA A.8.– *Consider a normed space (E, N). The function $d : E \times E \to \mathbb{R}_+$ defined by*

$$d(x, y) = N(x - y),$$

is a metric. In other words, any normed space is also a metric space.

A.3.5. *Hilbert spaces*

DEFINITION A.11.– *A metric space (E, d) is complete if any Cauchy sequence is convergent. In other words, if $\{x_n, n \geq 1\}$ is a sequence in E such that*

$$\lim_{n,m\to\infty} d(x_n, x_m) = 0,$$

then there exists $x \in E$ such that $\lim_{n\to\infty} d(x_n, x) = 0$.

DEFINITION A.12.– *A real inner product space which is also a complete metric space with respect to the metric induced by its inner product is called Hilbert space.*

The proof of the following well-known theorem can be found in many sources, see e.g. [BOU 04] or [DRI 04]. It is a partial case of the statement that the space $L^2(\Omega)$ is a Hilbert space.

THEOREM A.1.– *Consider* $(\Omega, \mathcal{F}, \mathbb{P})$ *a probability space. Then,* $L^2(\Omega) = L^2(\Omega, \mathcal{F}, \mathbb{P})$*, the space of square-integrable real-valued random variables* $X : \Omega \to \mathbb{R}$*, is a Hilbert space.*

LEMMA A.9.– *Consider two real inner product spaces* H *and* K*, and a linear isometry* $f : H \to K$*. If* H *is a Hilbert space, then* $f(H)$ *is also a Hilbert space.*

PROOF.– Let us denote by $\|\cdot\|_H$ (respectively, by $\|\cdot\|_K$) the norm induced by the inner product in H (respectively, in K). Consider a Cauchy sequence $\{y_n, n \geq 1\} = \{f(x_n), n \geq 1\}$ in $f(H)$. That is, $\lim_{n,m\to\infty} \|f(x_n) - f(x_m)\|_K = 0$. Because f is a linear isometry, we have

$$\|f(x_n) - f(x_m)\|_K = \|f(x_n - x_m)\|_K = \|x_n - x_m\|_H,$$

the sequence $\{x_n, n \geq 1\}$ is a Cauchy sequence in H. Because H is a Hilbert space, we deduce that there exists $x \in H$ such that

$$\lim_{n\to\infty} \|x_n - x\| = 0,$$

which implies that

$$\lim_{n\to\infty} \|f(x_n) - f(x)\| = \lim_{n\to\infty} \|f(x_n - x)\| = \lim_{n\to\infty} \|x_n - x\| = 0. \qquad \square$$

DEFINITION A.13.– *Consider a real inner product space* $(E, \langle\cdot,\cdot\rangle)$*. Two vectors* x *and* y *in* E *are said to be orthogonal if* $\langle x, y\rangle = 0$*. In such cases, we denote* $x \perp y$*. If* $A \subset E$ *is a subset of* E*, then* $x \in E$ *is orthogonal to* A *(we denote* $x \perp A$*) if* $x \perp y$ *for any* $y \in A$*.*

THEOREM A.2.– *Consider a Hilbert space* H*, and a closed subspace* $F \subset H$ *of* H*. Then, for any* $x \in H$*, there exists a unique* $y \in F$ *such that*

$$\|x - y\| = d(x, F) = \inf_{z\in F} \|x - z\| \text{ and } x - y \perp F.$$

Vector y *is called the orthogonal projection of* x *onto the space* F*.*

For the proof see, e.g. theorem 8.10, p. 78, in [DRI 04].

A.4. Elements of fractional calculus. Hardy–Littlewood theorem

Let $\alpha > 0$ and $f : \mathbb{R} \to \mathbb{R}$.

DEFINITION A.14.– *The Riemann–Liouville fractional integrals on $\mathbb{R}$ are defined by the equality*

$$(I_+^\alpha f)(x) := \frac{1}{\Gamma(\alpha)} \int_{-\infty}^{x} \frac{f(t)}{(x-t)^{1-\alpha}} dt,$$

and

$$(I_-^\alpha f)(x) := \frac{1}{\Gamma(\alpha)} \int_{x}^{+\infty} \frac{f(t)}{(t-x)^{1-\alpha}} dt,$$

respectively.

We say that the function $f \in \mathcal{D}(I_\pm^\alpha)$ if the corresponding integrals converge for a.a. $x \in \mathbb{R}$. According to [SAM 93], if $1 \leq p < \dfrac{1}{\alpha}$, then $L^p(\mathbb{R}, \lambda_1) \subset \mathcal{D}(I_\pm^\alpha)$, and moreover, we have the following theorem, known as the Hardy–Littlewood theorem.

THEOREM A.3.– *(theorem 5.3 in [SAM 93]) Let $1 \leq p \leq \infty$, $1 \leq q \leq \infty$, $\alpha > 0$. Operators $I_\pm^\alpha$ are bounded from $L^p(\mathbb{R}, \lambda_1)$ to $L^q(\mathbb{R}, \lambda_1)$ if and only if $0 < \alpha < 1$, $1 < p < \dfrac{1}{\alpha}$ and $q = \dfrac{p}{1-\alpha p}$. This implies, in particular, that for any $1 < p < \frac{1}{\alpha}$ and $q = \frac{p}{1-\alpha p}$, there exists a constant $C_{p,q,\alpha}$ such that*

$$\left(\int_{\mathbb{R}} \left(\int_{\mathbb{R}} |f(u)||x-u|^{\alpha-1} du \right)^q dx \right)^{\frac{1}{q}} \leq C_{p,q,\alpha} \|f\|_{L^p(\mathbb{R},\lambda_1)},$$

where λ_1 is the Lebesgue measure on $\mathbb{R}$.

A.5. Hausdorff and capacitarian dimensions

A.5.1. *Hausdorff measures and dimensions*

Let $d \geq 1$. Consider $E \subset \mathbb{R}^d$ and $\delta > 0$.

DEFINITION A.15.– *A countable collection of sets $\{E_k, k \geq 1\}$ is said to be a δ-cover of E, if $E \subset \cup_{k=1}^{\infty} E_k$ and $0 < |E_k| \leq \delta$ for each $k \geq 1$, where $|E_k|$ is the diameter of the set E_k defined by*

$$|E_k| = \sup\{|x-y| \; |x, y \in E_k\}.$$

Now, let additionally $\alpha > 0$.

DEFINITION A.16.– *The* α*-dimensional Hausdorff measure of* E *is defined by*

$$M^\alpha(E) = \lim_{\delta \to 0} \inf \left\{ \sum_{k=1}^{\infty} |E_k|^\alpha | \{E_k\}_{k=1}^{\infty} \text{ is a } \delta - \text{cover of } E \right\}. \qquad [A.2]$$

REMARK A.6.– The set function M^α is an outer measure on $\mathbb{R}^d$. It means the following.

i) $M^\alpha(E) \in [0, \infty]$ for any $E \in \mathcal{B}(\mathbb{R}^d)$.

ii) $M^\alpha(\emptyset) = 0$.

iii) $M^\alpha(E) \subset M^\alpha(F)$ if $E, F \in \mathcal{B}(\mathbb{R}^d)$ and $E \subset F$.

iv) $M^\alpha(\bigcup_{i\in\mathbb{N}} E_i) \leq \sum_{i\in\mathbb{N}} M^\alpha(E_i)$ if $E_i \in \mathcal{B}(\mathbb{R}^d)$ for any $i \in \mathbb{N}$.

The first assertion is a straightforward consequence of definition A.16. The proof of the three other assertions can be found in e.g. theorem 2.6, p. 701, in [YEH 14].

REMARK A.7.– It follows from the relation [A.2] that for any set E and $\delta < 1$,

$$\inf \left\{ \sum_{k=1}^{\infty} |E_k|^s | \{E_k\}_{k=1}^{\infty} \text{ is a } \delta - \text{cover of } E \right\}$$

is non-increasing in s, so $M^s(E)$ is also non-increasing in s. Moreover, if $t > s$ and $\{E_k\}_{k=1}^{\infty}$ is a δ-cover of E, we have

$$\sum_{k=1}^{\infty} |E_k|^t \leq \delta^{t-s} \sum_{k=1}^{\infty} |E_k|^s.$$

Therefore, $\inf \left\{ \sum_{k=1}^{\infty} |E_k|^t | \{E_k\}_{k=1}^{\infty} \text{ is a } \delta - \text{cover of } E \right\}$

$$\leq \delta^{t-s} \inf \left\{ \sum_{k=1}^{\infty} |E_k|^s | \{E_k\}_{k=1}^{\infty} \text{ is a } \delta - \text{cover of } E \right\},$$

which implies that $\lim_{\delta \to 0} \inf \left\{ \sum_{k=1}^{\infty} |E_k|^t | \{E_k\}_{k=1}^{\infty} \text{ is a } \delta - \text{cover of } E \right\}$

$$\leq \lim_{\delta \to 0} \delta^{t-s} \inf \left\{ \sum_{k=1}^{\infty} |E_k|^s | \{E_k\}_{k=1}^{\infty} \text{ is a } \delta - \text{cover of } E \right\}.$$

Hence, if $M^s(E) < \infty$, then $M^t(E) = 0$ for any $t > s$. Thus, a graph of $M^s(E)$ versus s shows that there is a critical value s_0 of s at which $M^s(E)$ jumps from ∞ to 0: $M^s(E) = \infty$ for any $s < s_0$ and $M^s(E) = 0$ for any $s > s_0$. This critical value is called the Hausdorff dimension of E, and it is defined as follows:

DEFINITION A.17.– *The Hausdorff dimension of E is defined by*

$$\dim E = \inf\{\alpha > 0 | M^\alpha(E) = 0\} = \sup\{\alpha > 0 | M^\alpha(E) = +\infty\}. \qquad [A.3]$$

REMARK A.8.–

i) From remark A.7 and equality [A.3], we obtain that

$$M^s(E) = \begin{cases} \infty \text{ if } s < \dim E \\ 0 \text{ if } s > \dim E. \end{cases}$$

ii) If $s = \dim E$, then $M^s(E)$ may be zero or infinite, or may satisfy

$$0 < M^s(E) < \infty.$$

In the following proposition, we give some properties of the Hausdorff dimension which are used in the book.

PROPOSITION A.1.–

i) Let $E \subset F$. Then, $\dim E \leq \dim F$.

ii) Let $\{F_i, i \in \mathbb{N}\}$ be a countable sequence of sets. Then, $\dim \bigcup_{i=0}^{\infty} F_i = \sup_{i\in\mathbb{N}} \dim F_i$.

PROOF.–

1) This is a straightforward consequence of the fact that $M^s(E) \leq M^s(F)$, for any s.

2) For any $i \in \mathbb{N}$, $F_i \subset \bigcup_{i=0}^{\infty} F_i$. Thus, it follows from assertion 1 of this proposition that $\dim F_i \leq \dim \bigcup_{i=0}^{\infty} F_i$ for any i, and consequently,

$$\sup_{i\in\mathbb{N}} \dim F_i \leq \dim \bigcup_{i=0}^{\infty} F_i.$$

Now, if $s > \sup_{i\in\mathbb{N}} \dim F_i$, then $M^s(F_i) = 0$ for any i. Moreover, from remark A.6, we have that $M^s(\bigcup_{i=0}^{\infty} F_i) \le \sum_{i=0}^{\infty} M^s(F_i)$. Therefore, $M^s(\bigcup_{i=0}^{\infty} F_i) = 0$, and consequently, $s \ge \dim \bigcup_{i=0}^{\infty} F_i$. It means that

$$(\sup_{i\in\mathbb{N}} \dim F_i, +\infty) \subset (\dim \bigcup_{i=0}^{\infty} F_i, +\infty),$$

whence the second inequality follows. □

COROLLARY A.2.– *If F is a Borel set in $\mathbb{R}^d$ with non-empty interior $\overset{\circ}{F} \neq \emptyset$, then $\dim F = d$.*

For the proof see, e.g. theorem 2.3, p. 717, in [YEH 14].

LEMMA A.10.– *Let $[a, b] \subset \mathbb{R}$ and $f : [a, b] \to \mathbb{R}$ be a Hölder continuous function of order α. That is, for any $x, y \in [a, b]$ $|f(x) - f(y)| \le c|x - y|^{\alpha}$, where $c > 0$ and $0 < \alpha \le 1$ are the constants. Then,*

$$\dim(Grf([a, b])) := \dim\{(t, f(t)), t \in [a, b]\} \le \min\left(\frac{1}{\alpha}, 2 - \alpha\right).$$

The proof of lemma A.10 can be found in e.g. [ADL 81], p. 193.

A.5.2. *Energy, capacity and Frostman's theorem*

Consider $0 < \beta < d$.

DEFINITION A.18.– *Given μ a signed measure, the energy integral of μ with respect to the kernel $|x|^{-\beta}$ is $I_\beta(\mu) = \int_{\mathbb{R}^d} \int_{\mathbb{R}^d} \frac{d\mu(x)d\mu(y)}{|x - y|^{\beta}}$. A signed measure μ is said to have finite β-energy if $I_\beta(\mu) < +\infty$.*

The potential of μ with respect to the kernel $|x|^{-\beta}$ is $U_\beta^\mu(x) := \int_{\mathbb{R}^d} \frac{d\mu(y)}{|x - y|^{\beta}}$.

LEMMA A.11.– *If we denote $k_\beta(x) = |x|^{-\beta}$, then k_β is locally integrable on $\mathbb{R}^d$ and*

$$k_\beta = C(\beta)\ k_{\beta/2} \star k_{\beta/2}$$

where $C(\beta)$ is a positive constant depending only on β, and

$$k_{\beta/2} \star k_{\beta/2}(x) := \int_{\mathbb{R}^d} k_{\beta/2}(x - y) k_{\beta/2}(y) dy.$$

The proof of lemma A.11 can be found in e.g. [LAN 72].

LEMMA A.12.– *If the potential U_β^μ of a signed measure μ is equal to zero a. e., then $\mu \equiv 0$.*

For the proof of lemma A.12 see, e.g. [LAN 72]. From lemmas A.11 and A.12, we get the following result.

COROLLARY A.3.– *For any signed measure μ, we have $I_\beta(\mu) \geq 0$. Moreover, $I_\beta(\mu) = 0$ if and only if $\mu \equiv 0$.*

PROOF.– It follows from lemma A.11 that

$$\begin{aligned}
I_\beta(\mu) &= \int_{\mathbb{R}^d}\int_{\mathbb{R}^d} k_\beta(x-y)d\mu(x)d\mu(y) \\
&= C(\beta)\int_{\mathbb{R}^d}\int_{\mathbb{R}^d}\left(\int_{\mathbb{R}^d} k_{\beta/2}(x-y-z)k_{\beta/2}(z)dz\right)d\mu(x)d\mu(y) \\
&= C(\beta)\int_{\mathbb{R}^d}\int_{\mathbb{R}^d}\left(\int_{\mathbb{R}^d} k_{\beta/2}(x-z)k_{\beta/2}(z-y)dz\right)d\mu(x)d\mu(y) \\
&= C(\beta)\int_{\mathbb{R}^d}\left(\int_{\mathbb{R}^d} k_{\beta/2}(x-z)d\mu(x)\int_{\mathbb{R}^d} k_{\beta/2}(z-y)d\mu(y)\right)dz \\
&= C(\beta)\int_{\mathbb{R}^d}\left(\int_{\mathbb{R}^d} k_{\beta/2}(x-z)d\mu(x)\right)^2 dz \geq 0.
\end{aligned}$$

Now, let $I_\beta(\mu) = 0$. Then, $\int_{\mathbb{R}^d} k_{\beta/2}(x-z)d\mu(x) = U_{\beta/2}^\mu(z) = 0$ a. e. From lemma A.12, we deduce that $\mu \equiv 0$. □

DEFINITION A.19.– *The mutual energy of signed measures μ_1 and μ_2 with respect to the kernel $|x|^{-\beta}$ is a bilinear functional*

$$I_\beta(\mu_1,\mu_2) = \int_{\mathbb{R}^d}\int_{\mathbb{R}^d} \frac{\mu_1(dt)\mu_2(ds)}{|t-s|^\beta}.$$

PROPOSITION A.2.– *For any signed measures μ and ν, denote $\langle \mu,\nu\rangle_\beta = I_\beta(\mu,\nu)$. Then, $\langle\cdot,\cdot\rangle$ is an inner product on the set of signed measures.*

PROOF.– The function $\langle \cdot, \cdot \rangle_\beta$ is obviously bilinear and symmetric. Moreover, from corollary A.3, for any signed measure μ, we have $\langle \mu, \mu \rangle_\beta = I_\beta(\mu) \geq 0$ and $\langle \mu, \mu \rangle_\beta = 0$ if and only if $\mu = 0$. That is, we have the positive-definiteness of $\langle \cdot, \cdot \rangle_\beta$. Consequently, $\langle \cdot, \cdot \rangle_\beta$ is an inner product. □

Denote $|\cdot|_\beta$ the associated norm to the inner product $\langle \cdot, \cdot \rangle_\beta$; that is, for a signed measure μ, we set $|\mu|_\beta = (\langle \mu, \mu \rangle_\beta)^{1/2}$.

COROLLARY A.4.– *The function $d : (\mu, \nu) \longmapsto I_\beta(\mu - \nu)$ is a metric on the space of signed measures.*

PROOF.– It suffices to remark that d is the metric associated to the norm $|\cdot|_\beta$. That is, for every measures μ and ν,

$$d(\mu, \nu) = |\mu - \nu|_\beta.$$

□

For a compact set K in $\mathbb{R}^d$, denote $M^+(K)$ the set of non-zero positive measures carried by K, $M_1^+(K)$ the subset of $M^+(K)$, which consists of probability measures. Also, denote $\mathcal{M}_\beta^+$ the space of all non-negative measures on $\mathbb{R}$ with finite β-energy.

The proof of the following theorem can be found in [LAN 72], p. 90.

THEOREM A.4.– *Let $\beta > 0$. Then, $\mathcal{M}_\beta^+$ is a complete metric space under the metric defined in corollary A.4.*

REMARK A.9.– Theorem A.4 means that if $\{\mu_i, i \geq 1\}$ is a Cauchy sequence in $\mathcal{M}_\beta^+$ (i.e. $d(\mu_m, \mu_n) = |\mu_m - \mu_n|_\beta \to 0$ as $m, n \to \infty$), then there exists a limit measure $\mu \in \mathcal{M}_\beta^+$ such that $|\mu_n - \mu|_\beta \to 0$ as $n \to \infty$.

DEFINITION A.20.– *We say that a signed measure μ is carried by a set K if $\mu(\mathbb{R}^d) = \mu(K)$. A signed measure μ, carried by a compact set K, is strictly positive if $0 < \mu(K) = \mu(\mathbb{R}^d) < \infty$.*

DEFINITION A.21.– *We say that K has a positive capacity with respect to the kernel $|x|^{-\beta}$ and we write $\mathrm{Cap}_\beta \mathrm{K} > 0$, if K carries strictly positive measures of finite β-energy. We say that K has capacity zero, with respect to the kernel $|x|^{-\beta}$, and we write $\mathrm{Cap}_\beta(\mathrm{K}) = 0$, if $I_\beta(\mu) = +\infty$ for any positive measure μ carried by K.*

REMARK A.10.– We have $\sup\{\beta | \mathrm{Cap}_\beta(\mathrm{K}) > 0\} = \inf\{\alpha | \mathrm{Cap}_\alpha(\mathrm{K}) = 0\}$. Indeed, if $\mathrm{Cap}_\beta \mathrm{K} > 0$, then there exists a positive measure μ carried by K such that

$$I_\beta(\mu) = \int_K \int_K \frac{d\mu(x) d\mu(y)}{|x - y|^\beta} < +\infty.$$

This implies that $\alpha > \beta$ for any $\alpha > 0$ such that $\mathrm{Cap}_\alpha \mathrm{K} = 0$. Because otherwise $\alpha \leq \beta$ would exist such that for any positive measure ν carried by K, we have

$$I_\alpha(\nu) = \int_K \int_K \frac{d\nu(x)d\nu(y)}{|x-y|^\alpha} = +\infty,$$

which would imply that there exists a positive constant C, depending only on K, α and β such that

$$I_\beta(\mu) = \int_K \int_K \frac{|x-y|^{\alpha-\beta} d\mu(x)d\mu(y)}{|x-y|^\alpha} \geq C \int_K \int_K \frac{d\mu(x)d\mu(y)}{|x-y|^\alpha} = \infty,$$

which is in contradiction to $I_\beta(\mu) < \infty$. Thus,

$$\sup\{\beta \,|\mathrm{Cap}_\beta(\mathrm{K}) > 0\} \leq \inf\{\beta|\mathrm{Cap}_\beta(\mathrm{K}) = 0\}.$$

This relation immediately implies that

$$\sup\{\beta \,|\mathrm{Cap}_\beta(\mathrm{K}) > 0\} = \inf\{\beta|\mathrm{Cap}_\beta(\mathrm{K}) = 0\}.$$

Indeed, otherwise, we would have

$$\sup\{\beta \,|\mathrm{Cap}_\beta(\mathrm{K}) > 0\} < \inf\{\beta|\mathrm{Cap}_\beta(\mathrm{K}) = 0\},$$

and consequently λ would exist such that

$$\sup\{\beta \,|\mathrm{Cap}_\beta(\mathrm{K}) > 0\} < \lambda < \inf\{\beta|\mathrm{Cap}_\beta(\mathrm{K}) = 0\},$$

which would imply that $\mathrm{Cap}_\lambda > 0$ and $\mathrm{Cap}_\lambda = 0$ at the same time, which is impossible. Thus,

$$\sup\{\beta \,|\mathrm{Cap}_\beta(\mathrm{K}) > 0\} = \inf\{\beta|\mathrm{Cap}_\beta(\mathrm{K}) = 0\}.$$

DEFINITION A.22.– *We call the capacitarian dimension of K the common value*

$$\mathrm{CapK} = \sup\{\beta \,|\mathrm{Cap}_\beta(\mathrm{K}) > 0\} = \inf\{\beta \,|\mathrm{Cap}_\beta(\mathrm{K}) = 0\}.$$

The following very important theorem is called Frostman's theorem.

THEOREM A.5.– *Consider K a compact set in $\mathbb{R}^d$, and $0 < \alpha < \beta < d$. Then,*

$$M^\beta(K) > 0 \Longrightarrow \mathrm{Cap}_\alpha \mathrm{K} > 0 \Longrightarrow \mathrm{M}^\alpha(\mathrm{K}) > 0.$$

For the proof of theorem A.5, see e.g. [KAH 85], p. 133.

As a consequence of theorem A.5, we get the following statement.

COROLLARY A.5.– *Consider a compact set K in $\mathbb{R}^d$. Suppose that there exists a strictly positive measure carried by K and a finite $\beta > 0$ such that the energy integrals $I_\alpha(K)$ are finite for any $\alpha < \beta$. Then,* $\dim K \geq \beta$.

PROOF.– From the hypothesis, K carries a strictly positive measure of finite α-energy, for any $\alpha < \beta$. This means that we have $\mathrm{Cap}_\alpha(\mathrm{K}) > 0$ for any $\alpha < \beta$. Thus, it follows from theorem A.5 that we have $M^\alpha(K) > 0$ for any $\alpha < \beta$. Consequently,

$$\dim K = \sup\{\alpha | M^\alpha(K) > 0\} \geq \beta. \qquad \square$$

REMARK A.11.– From corollary A.5, we see that to show that $\dim K \geq \beta$, it is enough to prove that there exists a strictly positive measure carried by K such that the energy integrals $I_\alpha(K)$ are finite for any $\alpha < \beta$.

A.6. On the total variation of a signed measure

The proof of the following theorem can be found in e.g. [DUD 02].

THEOREM A.6.– *For any signed measure μ on a measurable space $(E, \mathcal{B})$, there is a set $D \in \mathcal{B}$ such that for all $E \in \mathcal{B}$,*

$$\mu^+(E) := \mu(E \cap D) \geq 0 \text{ and } \mu^-(E) := -\mu(E \setminus D) \geq 0.$$

Then, μ^+ and μ^- are measures, at least one of which is finite, $\mu = \mu^+ - \mu^-$ and there exists a measurable set A such that $\mu^+(A) = \mu^-(E \setminus A) = 0$. These properties uniquely determine μ^+ and μ^-.

DEFINITION A.23.– *The relation $\mu = \mu^+ - \mu^-$ introduced in the theorem above is called the Jordan decomposition of μ. The measure $|\mu| := \mu^+ + \mu^-$ is called the total variation measure for μ.*

A.7. Elements of matrix analysis

A.7.1. *Notations and elementary definitions*

For $(n, p) \in \mathbb{N} \times \mathbb{N}$, we denote $\mathcal{M}_{n \times p}(\mathbb{R})$ the set of matrices with real coefficients and with n rows and p columns. That is,

$$\mathcal{M}_{n\times p}(\mathbb{R}) = \Big\{ (a_{ij})_{\substack{1\leq i\leq n \\ 1\leq j\leq p}} | a_{ij} \in \mathbb{R},\ 1 \leq i \leq n,\ \ 1 \leq j \leq p \Big\}.$$

If $n = p$, then the set $\mathcal{M}_{n\times n}(\mathbb{R})$ is also denoted $\mathcal{M}_n(\mathbb{R})$. In the particular case where $n = p$ and

$$a_{ij} = \begin{cases} 1 \text{ if } i = j \\ 0 \text{ if } i \neq j, \end{cases}$$

the matrix is denoted by I_n and called the identity matrix (or unit matrix).

For $M = (a_{ij})_{\substack{1\le i\le n,\\ 1\le j\le p}}$ we denote by M^T the transpose of M. That is,

$$M^T = (b_{ij})_{\substack{1\le i\le p\\ 1\le j\le n}} \in \mathcal{M}_{p\times n}(\mathbb{R})$$

where $b_{ij} = a_{ji}$ for any $1 \le i \le p \quad 1 \le j \le n$. A matrix $M \in \mathcal{M}_n(\mathbb{R})$ is called symmetric if $M^T = M$. A matrix $M = (a_{ij})_{1\le i,j\le n}$ is called diagonal, if $a_{ij} = 0$ whenever $j \neq i$.

DEFINITION A.24.– *A matrix $M \in \mathcal{M}_n(\mathbb{R})$ is called invertible (or non-singular) if there exists a matrix, which is denoted by M^{-1}, such that $MM^{-1} = M^{-1}M = I_n$. The matrix M^{-1} is called the inverse of M.*

DEFINITION A.25.– *A matrix $M \in \mathcal{M}_n(\mathbb{R})$ is called positive semi-definite if*

$$X^T MX \ge 0$$

for any vector $X \in \mathbb{R}^n$, and is called positive definite if

$$X^T MX > 0$$

for any vector $X \in \mathbb{R}^n \setminus \{0\}$.

DEFINITION A.26.– *A matrix $M \in \mathcal{M}_n(\mathbb{R})$ is called normal if $M^T M = MM^T$. It means that M commutes with its transpose.*

DEFINITION A.27.– *A matrix $M \in \mathcal{M}_n(\mathbb{R})$ is called orthogonal if $M^T M = MM^T = I_n$. It means that M is invertible and $M^{-1} = M^T$.*

REMARK A.12.– We see that any orthogonal matrix is normal.

The following theorems belong to the extended group of theorems of simultaneous diagonal matrix diagonalization. Their proofs can be found in [HOR 13, SER 88, THO 91, LAN 05]. In particular, for theorem A.9, see corollary 7.6.5 from [HOR 13].

THEOREM A.7.– *Consider a symmetric matrix $A \in \mathcal{M}_n(\mathbb{R})$. Then, there exists an orthogonal matrix S such that $S^T AS$ is a diagonal matrix.*

THEOREM A.8.– *Let A and B be a pair of non-negative definite symmetric matrices with real elements. Then, there exists a non-degenerate matrix S such that*

$$S^T AS = \begin{bmatrix} I_r & 0 & 0 \\ 0 & 0 & 0 \\ 0 & 0 & 0 \end{bmatrix}, S^T BS = \begin{bmatrix} D_r & 0 & 0 \\ 0 & I_p & 0 \\ 0 & 0 & 0 \end{bmatrix}, D_r = \begin{bmatrix} \lambda_1 & & 0 \\ & \ddots & \\ 0 & & \lambda_r \end{bmatrix},$$

and the numbers $\lambda_i \geq 0$. The form of the transformed diagonal matrices $S^T AS$ and $S^T BS$ is unique, up to possible permutations of numbers λ_i.

THEOREM A.9.– *Consider two symmetric matrices $A, B \in \mathcal{M}_n(\mathbb{R})$. If A is positive definite, then there exists an invertible matrix S such that $S^T AS = I_n$ and $S^T BS$ is a diagonal matrix.*

A.7.2. *Inequalities with eigenvalues*

DEFINITION A.28.– *A number $\lambda \in \mathbb{C}$ is called the eigenvalue of matrix $A \in \mathcal{M}_n(\mathbb{R})$ if there exists a vector $x \in \mathbb{R}^n \setminus \{0\}$ such that $Ax = \lambda x$.*

The proof of the following theorem can be found in [HOR 13], theorem 5.6.9, p. 347.

THEOREM A.10.– *Consider a matrix $A = (a_{ij})_{1 \leq i,j \leq n} \in \mathcal{M}_n(\mathbb{R})$. Then, any eigenvalue λ of A satisfies*

$$|\lambda| \leq \max_{1 \leq i \leq n} \sum_{j=1}^{n} |a_{ij}|.$$

Because the eigenvalues of any symmetric matrix $A \in \mathcal{M}_n(\mathbb{R})$ are real, we may adopt the convention that they are always arranged in decreasing order. In the subsequent part of this section, for symmetric matrix $A \in \mathcal{M}_n(\mathbb{R})$, we shall use the notation $\lambda_k(A)$ to designate the kth largest eigenvalue. Thus,

$$\lambda_n(A) \leq \cdots \leq \lambda_2(A) \leq \lambda_1(A).$$

The following theorem is due to e.g. [GOL 83], theorem 8.1.5.

THEOREM A.11.– *Consider two symmetric matrices A_1 and A_2 from $\mathcal{M}_n(\mathbb{R})$. Then, for any $k \in \{1, \ldots, n\}$,*

$$\lambda_k(A_1) + \lambda_n(A_2) \leq \lambda_k(A_1 + A_2) \leq \lambda_k(A_1) + \lambda_1(A_2).$$

COROLLARY A.6.– *Consider $r \geq 2$, and let A_1, A_2,..., A_r be r symmetric matrices from $\mathcal{M}_n(\mathbb{R})$. Then,*

$$\lambda_k\Big(\sum_{i=1}^{r} a_i A_i\Big) \leq \sum_{i=1}^{r} a_i \lambda_1(A_i), \qquad [A.4]$$

for any $1 \leq k \leq n$ and any $a_i \geq 0$.

PROOF.– Let us argue applying induction on r. For $r = 2$, we get from theorem A.11 that

$$\lambda_k(a_1 A_1 + a_2 A_2) \leq \lambda_k(a_1 A_1) + \lambda_1(a_2 A_2).$$

Because $\lambda_k(a_1 A_1) = a_1 \lambda_k(A_1) \leq a_1 \lambda_1(A_1)$ and $\lambda_1(a_2 A_2) = a_2 \lambda_1(A_2)$, we get inequality [A.4].

Let us suppose that inequality [A.4] is true for an integer $r \geq 2$, and let us check it for $r + 1$. We can deduce from theorem A.11 that

$$\lambda_k\Big(\sum_{i=1}^{r+1} a_i A_i\Big) \leq \lambda_k\Big(\sum_{i=1}^{r} a_i A_i\Big) + \lambda_1(a_{r+1} A_{r+1})$$
$$= \lambda_k\Big(\sum_{i=1}^{r} a_i A_i\Big) + a_{r+1} \lambda_1(A_{r+1}).$$

Thus, by the induction hypothesis, we get inequality [A.4]. □

The proof of the following theorem can be found in e.g. [HOR 13], theorem 4.2, p. 234.

THEOREM A.12.– *Consider a symmetric matrix $A \in \mathcal{M}_n(\mathbb{R})$ with real eigenvalues*

$$\lambda_n(A) \leq \cdots \leq \lambda_2(A) \leq \lambda_1(A).$$

Then,

$$\lambda_1(A) = \max_{x \neq 0} \frac{x^T A x}{x^T x} \quad \text{and} \quad \lambda_n(A) = \min_{x \neq 0} \frac{x^T A x}{x^T x}.$$

A.8. Elements of stochastic processes

A.8.1. *Filtrations*

Consider $(\Omega, \mathcal{F}, \mathbb{P})$ a probability space.

DEFINITION A.29.– *A filtration is a non-decreasing family* $\{\mathcal{F}_t, t \geq 0\}$ *of sub-*σ*–fields of* $\mathcal{F}$ *such that*

$$\mathcal{F}_s \subset \mathcal{F}_t \subset \mathcal{F} \text{ for } 0 \leq s < t < \infty.$$

We set $\mathcal{F}_\infty = \sigma\Big(\bigcup_{t\geq 0} \mathcal{F}_t\Big)$, *the sigma-field generated by the union* $\bigcup_{t\geq 0} \mathcal{F}_t$.

EXAMPLE A.7.– *Given a stochastic process* $X = \{X_t, t \geq 0\}$, *we can consider the filtration* $\{\mathcal{F}_t^X, t \geq 0\}$ *generated by the process itself, defined by:*

$$\mathcal{F}_t^X = \sigma(X_s, 0 \leq s \leq t),$$

the smallest σ*-field with respect to which any vector* $(X_{s_i}, 1 \leq i \leq p)$ *is measurable for any* $s_i \in [0, t], p \geq 1$. *This filtration is called the natural filtration of* X.

DEFINITION A.30.– *A filtration* $\{\mathcal{F}_t, t \geq 0\}$ *is right-continuous if* $\mathcal{F}_t = \mathcal{F}_{t^+}$ *for any* $t \geq 0$, *where*

$$\mathcal{F}_{t^+} = \bigcap_{\epsilon > 0} \mathcal{F}_{t+\epsilon}.$$

DEFINITION A.31.– *A filtration* $\{\mathcal{F}_t, t \geq 0\}$ *is said to satisfy the standard assumptions if it is right-continuous and* $\mathcal{F}_0$ *contains all the* $\mathbb{P}$*-negligible events from* $\mathcal{F}$. *For a filtration* $\mathcal{F} = \{\mathcal{F}_t, t \geq 0\}$, *we denote by* $\overline{\mathcal{F}} = \{\overline{\mathcal{F}}_t\}$ *the smallest filtration containing* $\mathcal{F}$ *and that satisfies the standard assumptions.*

A.8.2. *Variation and quadratic variation of stochastic process*

Let $X = \{X_t, t \geq 0\}$ be a stochastic process on a probability space $(\Omega, \mathcal{F}, \mathbb{P})$. Fix $t > 0$ and consider $\pi_m = \{t_0, t_1, \ldots, t_m\}$, with $0 = t_0 < t_1 < \cdots < t_m = t$, to be a partition of $[0, t]$.

DEFINITION A.32.– *The stochastic process is said to be of finite variation if its trajectories are a.s. of finite variation.*

DEFINITION A.33.– *The sum*

$$V_t^{(2)}(\pi_m) = \sum_{k=1}^{m} |X_{t_k} - X_{t_{k-1}}|^2$$

is called a quadratic variation of X over the partition π_m. If $V_t^{(2)}(\pi_m)$ converges in probability as $\|\pi_m\| \to 0$, and the limit does not depend on sequence π_m, then the limit is called the quadratic variation of X on $[0,t]$, and we denote it $[X]_t$. That is,

$$[X]_t = \lim_{\|\pi_m\| \to 0} \sum_{k=1}^{m} (X_{t_k} - X_{t_{k-1}})^2$$

in probability if the limit does not depend on π_m.

LEMMA A.13.– *If X is a continuous process of finite variation, then its quadratic variation is zero.*

PROOF.– Fix $t > 0$ and consider a partition $\pi_m = \{t_0, t_1, \dots, t_m\}$ of $[0,t]$. We have

$$V_t^{(2)}(\pi_m) = \sum_{k=1}^{m} |X_{t_k} - X_{t_{k-1}}|^2 \leq \sup_{1 \leq k \leq m} |X_{t_k} - X_{t_{k-1}}| V_t(X),$$

which tends to zero when $\|\pi_m\| \to 0$, because of the continuity of X and the finiteness of $V_t(X)$. □

A.8.3. *Orthogonal random measure and Karhunen theorem*

DEFINITION A.34.– *A complex-valued centered random function*

$$\mathcal{Z} = \mathcal{Z}(A, \omega) : \mathcal{B}(\mathbb{R}) \times \Omega \to \mathbb{C}$$

is called an orthogonal random measure with spectral measure μ on $\mathcal{B}(\mathbb{R})$ if it satisfies the following assumptions

1) $\mathbb{E}|\mathcal{Z}(A)|^2 < \infty$ for any $A \in \mathcal{B}(\mathbb{R})$.

2) $\mathbb{E}\Big(\mathcal{Z}(A)\overline{\mathcal{Z}(A)}\Big) = \mu(A \cap B)$ for any $A, B \in \mathcal{B}(\mathbb{R})$,

where μ is a σ-finite measure on $\mathcal{B}(\mathbb{R})$.

REMARK A.13.– Let $\mathcal{Z}$ be an orthogonal random measure with spectral measure μ of the form $\mu(A) = \int_A \varphi(x) d\lambda_1(x)$, where $\varphi \geq 0$ and $\varphi \in L^1(\mathbb{R})$. Then,

$$\mathcal{Z}(A) = \int_A \sqrt{\varphi(x)} W(dx),$$

where W is a Wiener measure on $\mathbb{R}$, i.e. $W(A)$ and $W(B)$ are independent for any Borel sets A, B; $A \cap B = \emptyset$, $W(A)$ is a Gaussian random variable with $\mathbb{E}(W(A)) = 0$ and $\mathbb{E}(W^2(A)) = \lambda_1(A)$.

Now we formulate the following version of the Karhunen theorem.

THEOREM A.13.– *Let $X = \{X_t, t \in \mathbb{T}\}$ be a complex-valued square-integrable centered stochastic process whose covariance function admits the following representation*

$$\mathbb{E}(X_s \overline{X_t}) = \int_{\mathbb{R}} f(s, \lambda)\overline{f(t, \lambda)}\mu(d\lambda), \ s, t \in \mathbb{T},$$

where $\int_{\mathbb{R}} |f(s, \lambda)|^2 \mu(d\lambda) < \infty$, for any $s \in \mathbb{T}$, and μ is a σ -finite measure on $\mathcal{B}(\mathbb{R})$. Then, there exists an orthogonal random measure $\mathcal{Z}$ (possibly on some extension of the initial probability space) such that its spectral measure is μ and

$$X_s = \int_{\mathbb{R}} f(s, x)\mathcal{Z}(dx), \ s \in \mathbb{T}.$$

A.8.4. *Characteristic functions of finite-dimensional distributions of fractional process defined via its spectral representation*

Consider a stochastic process $Y = \{Y_t, t \geq 0\}$ defined with the help of the spectral representation:

$$Y_t = \int_{\mathbb{R}} \frac{\exp(itx) - 1}{ix} |x|^{\frac{1}{2} - H} dW(x), \qquad [A.5]$$

where $0 < H < 1$, where W is a Wiener measure whose frequency domain is $\{\mathbb{R}\}$ and such that $\mathbb{E}W(A) = 0$ for any Borel set A, $W(A)$ and $W(B)$ are independent for $A \cap B = \emptyset$, $W(A) = W(-A)$, and

$$\mathbb{E}W(A)^2 = \lambda_1(A),$$

where λ_1 denotes the Lebesgue measure on $\mathbb{R}$. The proof of the following Lemma can be found in e.g. [SAM 94], section 7.7.

LEMMA A.14.– *Stochastic process Y is well defined, and the characteristic function of any of its finite-dimensional distribution is given by the formula*

$$\mathbb{E}\exp\left(i\sum_{j=1}^{n} \theta_j Y_{t_j}\right) = \exp\left(-\frac{1}{2}\int_{\mathbb{R}} \left|\sum_{j=1}^{n} \theta_j \frac{e^{ixt_j} - 1}{ix} x^{\frac{1}{2} - H}\right|^2 dx\right),$$

$n \geq 1$, $t_1, \ldots, t_n \in \mathbb{R}_+$ and $\theta = (\theta_1, \ldots, \theta_n) \in \mathbb{R}^n$.

A.9. Two Fernique's theorems

The following two theorems are known as Fernique's zero-one law and Fernique's theorem, respectively. The proof of the zero-one law can be found in [FER 75], corollary 1.3, p. 9.

THEOREM A.14.– *Fernique's zero-one law. Consider a continuous Gaussian process* $X = (X_t)_{t\in\mathbb{T}}$ *and a pseudo-semi-norm* N *on* $C(\mathbb{T})$*. Then, we have the alternatives:*

1) $\mathbb{P}(N(X) < \infty) = 0$ *or* $\mathbb{P}(N(X) < \infty) = 1$.

2) $\mathbb{P}(N(X) = 0) = 0$ *or* $\mathbb{P}(N(X) = 0) = 1$.

The following statement, called Fernique's theorem, was first established in [FER 70], in a more simple situation, and for the general result, see [FER 75], theorem 1.3, p. 11.

THEOREM A.15.– *Consider a continuous Gaussian process* $X = \{X_t, t \in \mathbb{T}\}$*, where* $\mathbb{T} \subset \mathbb{R}$*, and a pseudo-semi-norm* N *on* $C(\mathbb{T})$*. Suppose that* $\mathbb{P}(N(X) < \infty) > 0$*. Then, there exists* $\epsilon > 0$ *such that for any* $\alpha < \epsilon$

$$\mathbb{E}\exp(\alpha N^2(X)) < \infty.$$

COROLLARY A.7.– *Consider a continuous Gaussian process* $X = \{X_t, t \in \mathbb{T}\}$*, where* $\mathbb{T} \subset \mathbb{R}$*. Then, there exists* $\epsilon > 0$ *such that for any* $\alpha < \epsilon$

$$\mathbb{E}\exp\left(\alpha \sup_{t\in\mathbb{T}} X_t^2\right) < \infty,$$

and consequently for any $k \in \mathbb{N}$

$$\mathbb{E}\sup_{t\in\mathbb{T}} X_t^{2k} < \infty.$$

A.10. Weak convergence of stochastic processes

DEFINITION A.35.– *Consider a sequence* $\{X_n, n \in \mathbb{N}\}$ *of random variables defined on a probability space* $(\Omega, \mathcal{F}, \mathbb{P})$*, with values in a metric space* (S, d)*. The sequence* $\{X_n, n \in \mathbb{N}\}$ *is called tight if, for every* $\epsilon > 0$*, there is a compact set* $K \subset S$ *such that* $\mathbb{P}(X_n \in K) \geq 1 - \epsilon$*, for every* $n \in \mathbb{N}$.

DEFINITION A.36.– *Consider finite measures* μ *and* $\{\mu_n, n \geq 1\}$ *defined on a metric space, endowed with the Borel* σ*-field. The sequence* $\{\mu_n, n \geq 1\}$ *converges weakly to* μ*, and we denote* $\mu_n \rightsquigarrow \mu$*, if*

$$\int_S f(x)d\mu_n(x) \to \int_S f(x)d\mu(x) \text{ as } n \to \infty,$$

for every continuous and bounded function $f : S \to \mathbb{R}$.

Denote by $C([0,T])$ the set of all continuous real-valued functions on $[0,T]$ endowed with the supremum metric. Consider a sequence $\{X_n, n \geq 1\}$ of stochastic processes defined on a probability space $(\Omega, \mathcal{F}, \mathbb{P})$ with induced laws $\{\mu_n, n \geq 1\}$. Consider also X a stochastic process defined on $(\Omega, \mathcal{F}, \mathbb{P})$, with induced law μ. Suppose that the sample paths of the processes X and X_n, $n \in \mathbb{N}$, are continuous on $[0,T]$. Therefore, $\{X_n, n \geq 1\}$ (respectively, X) can be seen as a sequence of random variables (a random variable), namely,

$$X_n \ (resp.\ X) : \Omega \to C([0,T]).$$

It means that in this case $S = C([0,T])$ supplied with the sigma-field generated by cylinder sets.

DEFINITION A.37.– *The sequence $\{X_n, n \geq 1\}$ converges weakly to X, and we denote $X_n \rightsquigarrow X$ if $\mu_n \rightsquigarrow \mu$.*

The following theorem gives the link between weak convergence and tightness of a sequence of continuous stochastic process. Its proof can be found in e.g. [VAA 96], p. 35.

THEOREM A.16.– *Consider $(X_n)_{n \in \mathbb{N}}$ (respectively, X) a sequence of stochastic processes (a stochastic process) with sample paths in $C([0,T])$. Then, $X_n \rightsquigarrow X$ if and only if*

i) The finite-dimensional distributions (fdd's) of the processes X_n converge weakly to the fdd's of the process X.

ii) The sequence $\{X_n, n \geq 1\}$ is tight in $C([0,T])$.

The following theorem is known as the Lévy–Itô–Nisio theorem. Its proof can be found in e.g. [VAA 96], p. 431.

THEOREM A.17.– *Let $\{X_n, n \geq 1\}$ be a sequence of independent stochastic processes with sample paths in $C([0,T])$. Denote $S_n = \sum_{i=1}^{n} X_i$. Then, the following statements are equivalent.*

i) $\{S_n, n \geq 1\}$ converges in $C([0,T])$ with probability 1.

ii) $\{S_n, n \geq 1\}$ converges weakly in $C([0,T])$.

DEFINITION A.38.– *Consider a sequence* $\{X_n, n \geq 1\}$ *of stochastic processes with sample paths in* $C([0,T])$. *The sequence is called uniformly equicontinuous in probability if for any* $\epsilon, \eta > 0$ *there exists* $\delta > 0$ *such that*

$$\sup_{n\in\mathbb{N}} \mathbb{P}\left(\sup_{|t-s|<\delta} |X_n(t) - X_n(s)| > \epsilon \right) < \eta.$$

The proof of the following theorem can be found in e.g. [VAA 96], p. 37.

THEOREM A.18.– *Consider a sequence* $\{X_n, n \geq 1\}$ *of stochastic processes with sample paths in* $C([0,T])$. *The sequence is tight if and only if* $\{X_n, n \in \mathbb{N}\}$ *is equicontinuous in probability and for every* $t \in [0,T]$ *the sequence of random variables* $\{X_n(t), n \in \mathbb{N}\}$ *is tight in* $\mathbb{R}$. *The latter means that*

$$\limsup_{n\to\infty} \lim_{C\to\infty} \mathbb{P}\left(|X_n(t)| \geq C\right) = 0.$$

Bibliography

[ACH 10] ACHARD S., COEURJOLLY J.-F., "Discrete variations of the fractional Brownian motion in the presence of outliers and an additive noise", *Statistics Surveys*, vol. 4, pp. 117–147, 2010.

[ADL 81] ADLER R.J., *The Geometry of Random Fields*, Wiley Series in Probability and Mathematical Statistics, John Wiley & Sons, Ltd., Chichester, 1981.

[ALÒ 01] ALÒS E., MAZET O., NUALART D., "Stochastic calculus with respect to Gaussian processes", *Annals of Probability*, vol. 29, no. 2, pp. 766–801, 2001.

[AND 06] ANDROSHCHUK T., MISHURA Y., "Mixed Brownian–fractional Brownian model: absence of arbitrage and related topics", *Stochastics*, vol. 78, no. 5, pp. 281–300, 2006.

[AYA 00] AYACHE A., LEVY VEHEL J., "The generalized multifractional Brownian motion", *Statistical Inference for Stochastic Processes*, vol. 3, nos 1–2, pp. 7–18, 2000.

[AYA 05] AYACHE A., XIAO Y., "Asymptotic properties and Hausdorff dimensions of fractional Brownian sheets", *Journal of Fourier Analysis and Applications*, vol. 11, no. 4, pp. 407–439, 2005.

[BAL 08] BALAN R.M., TUDOR C.A., "The stochastic heat equation with fractional-colored noise: existence of the solution", *ALEA – Latin American Journal of Probability and Mathematical Statistics*, vol. 4, pp. 57–87, 2008.

[BEN 98] BENASSI A., COHEN S., ISTAS J., "Identifying the multifractional function of a Gaussian process", *Statistics & Probability Letters*, vol. 39, no. 4, pp. 337–345, 1998.

[BER 85] BERMAN S.M., "An asymptotic bound for the tail of the distribution of the maximum of a Gaussian process", *Ann. Inst. H. Poincaré Probab. Statist.*, vol. 21, no. 1, pp. 47–57, 1985.

[BIA 08] BIAGINI F., HU Y., ØKSENDAL B. *et al.*, *Stochastic Calculus for Fractional Brownian Motion and Applications*, Springer-Verlag, Ltd., London, 2008.

[BIE 09] BIERMÉ H., LACAUX C., XIAO Y., "Hitting probabilities and the Hausdorff dimension of the inverse images of anisotropic Gaussian random fields", *Bulletin of the London Mathematical Society*, vol. 41, no. 2, pp. 253–273, 2009.

[BOJ 04] BOJDECKI T., GOROSTIZA L.G., TALARCZYK A., "Sub-fractional Brownian motion and its relation to occupation times", *Statistics & Probability Letters*, vol. 69, no. 4, pp. 405–419, 2004.

[BOR 02] BORODIN A.N., SALMINEN P., *Handbook of Brownian Motion–Facts and Formulae*, 2nd edition, Probability and its Applications, Birkhäuser Verlag, Basel, 2002.

[BOR 17] BOROVKOV K., MISHURA Y., NOVIKOV A. *et al.*, "Bounds for expected maxima of Gaussian processes and their discrete approximations", *Stochastics*, vol. 89, no. 1, pp. 21–37, 2017.

[BOU 04] BOURBAKI N., *Integration. I. Chapters 1–6*, translated by BERBERIAN S.K., Elements of Mathematics, Springer-Verlag, Berlin, 2004.

[CAI 16] CAI C., CHIGANSKY P., KLEPTSYNA M., "Mixed Gaussian processes: a filtering approach", *Annals of Probability*, vol. 44, no. 4, pp. 3032–3075, 2016.

[CHE 01] CHERIDITO P., "Mixed fractional Brownian motion", *Bernoulli*, vol. 7, no. 6, pp. 913–934, 2001.

[CHE 03] CHERIDITO P., KAWAGUCHI H., MAEJIMA M., "Fractional Ornstein-Uhlenbeck processes", *Electronic Journal of Probability*, vol. 8, pp. 1–14, 2003.

[COE 05] COEURJOLLY J.-F., "Identification of multifractional Brownian motion", *Bernoulli*, vol. 11, no. 6, pp. 987–1008, 2005.

[DAL 99] DALANG R., "Extending martingale measure stochastic integral with applications to spatially homogeneous s.p.d.e.'s", *Electronic Journal of Probability*, vol. 4, no. 6, pp. 1–29, 1999.

[DEA 11] DEAN C.R., YOUNG A.F., CADDEN-ZIMANSKY P. *et al.*, "Multicomponent fractional quantum Hall effect in graphene", *Nature Physics*, vol. 7, no. 9, pp. 693–696, 2011.

[DEL 80] DELLACHERIE C., MEYER P.-A., *Probabilités et potentiel. Chapitres V à VIII*, Actualités Scientifiques et Industrielles, vol. 1385, revised edition, Hermann, Paris, 1980.

[DOM 11] DOMINIQUE C.-R., RIVERA-SOLIS L.E.S., "Mixed fractional Brownian motion, short and long-term Dependence and economic conditions: the case of the S&P-500 Index", *International Business and Management*, vol. 3, no. 2, pp. 1–6, 2011.

[DOO 53] DOOB J.L., *Stochastic Processes*, John Wiley & Sons, Inc., New York, 1953.

[DOU 02] DOUKHAN P., OPPENHEIM G., TAQQU M. (eds), *Theory and Applications of Long-range Dependence*, Springer Science & Business Media, 2002.

[DOZ 15] DOZZI M., MISHURA Y., SHEVCHENKO G., "Asymptotic behavior of mixed power variations and statistical estimation in mixed models", *Statistical Inference for Stochastic Processes*, vol. 18, no. 2, pp. 151–175, 2015.

[DRI 04] DRIVER B.K., Analysis tools with examples, 2004.

[DUD 02] DUDLEY R.M., *Real Analysis and Probability*, Cambridge Studies in Advanced Mathematics, vol. 74, Cambridge University Press, Cambridge, 2002.

[DZH 04] DZHAPARIDZE K., VAN ZANTEN H., "A series expansion of fractional Brownian motion", *Probablity Theory Related Fields*, vol. 130, no. 1, pp. 39–55, 2004.

[DZH 05] DZHAPARIDZE K., VAN ZANTEN H., "Krein's spectral theory and the Paley-Wiener expansion for fractional Brownian motion", *The Annnals of Probability*, vol. 33, no. 2, pp. 620–644, 2005.

[ELN 03] EL-NOUTY C., "The fractional mixed fractional Brownian motion", *Statistics & Probability Letters*, vol. 65, no. 2, pp. 111–120, 2003.

[ELN 15] EL-NOUTY C., ZILI M., "On the sub-mixed fractional Brownian motion", *Applied Mathematics – A Journal of Chinese Universities*, vol. 30, no. 1, pp. 27–43, 2015.

[ENG 80] ENGELBERT H.J., SHIRYAEV A.N., "On absolute continuity and singularity of probability measures", *Mathematical Statistics*, vol. 6, pp. 121–132, Banach Center Publisher, PWN, Warsaw, 1980.

[FEL 58] FELDMAN J., "Equivalence and perpendicularity of Gaussian processes", *Pacific Journal of Mathematics*, vol. 8, pp. 699–708, 1958.

[FEL 71] FELLER W., *An Introduction to Probability Theory and its Applications. Vol. II*, 2nd edition, John Wiley & Sons, Inc., 1971.

[FER 70] FERNIQUE X., "Intégrabilité des vecteurs gaussiens", *Comptes Rendus de l'Académie des Sciences*, vol. 270, pp. A1698–A1699, 1970.

[FER 75] FERNIQUE X., "Regularité des trajectoires des fonctions aléatoires Gaussiennes", in *École d'Été de Probabilités de Saint-Flour, IV-1974*, Lecture Notes in Mathematics, vol. 480, Springer, Berlin, 1975.

[FIL 08] FILATOVA D., "Mixed fractional Brownian motion: some related questions for computer network traffic modeling", *International Conference on Signals and Electronic Systems, Kraków 2008*, pp. 393–396, 2008.

[GOL 83] GOLUB G.H., VAN LOAN C.F., *Matrix Computations*, Johns Hopkins Series in the Mathematical Sciences, vol. 3, Johns Hopkins University Press, Baltimore, 1983.

[GRA 80] GRADSHTEYN I.S., RYZHIK I.M., *Table of Integrals, Series, and Products*, Academic Press, 1980.

[HÁJ 58] HÁJEK J., "On a property of normal distribution of any stochastic process", *Czechoslovak Mathematical Journal*, vol. 8, no. 83, pp. 610–618, 1958.

[HER 10] HERRMANN R., "Higher-dimensional mixed fractional rotation groups as a basis for dynamic symmetries generating the spectrum of the deformed Nilsson oscillator", *Journal of Physics A*, vol. 389, no. 4, pp. 693–704, 2010.

[HID 93] HIDA T., HITSUDA M., *Gaussian processes*, Translations of Mathematical Monographs, vol. 120, American Mathematical Society, Providence, 1993.

[HOR 13] HORN R.A., JOHNSON C.R., *Matrix Analysis*, 2nd edition, Cambridge University Press, Cambridge, 2013.

[HUR 51] HURST H.E., "Long-term storage capacity of reservoirs", *Transactions of the American Society of Civil Engineers*, vol. 116, pp. 400–410, 1951.

[JOS 06] JOST C., "Transformation formulas for fractional Brownian motion", *Stochastic Processes and their Applications*, vol. 116, no. 10, pp. 1341–1357, 2006.

[KAH 85] KAHANE J.-P., *Some random series of functions*, 2nd edition, Cambridge Studies in Advanced Mathematics, vol. 5, Cambridge University Press, Cambridge, 1985.

[KOL 40] KOLMOGOROFF A.N., "Wienersche Spiralen und einige andere interessante Kurven im Hilbertschen Raum", *C. R. (Doklady) Acad. Sci. URSS (N.S.)*, vol. 26, pp. 115–118, 1940.

[KOZ 15] KOZACHENKO Y., MELNIKOV A., MISHURA Y., "On drift parameter estimation in models with fractional Brownian motion", *Statistics*, vol. 49, no. 1, pp. 35–62, 2015.

[KUA 15a] KUANG N., LIU B., "Parameter estimations for the sub-fractional Brownian motion with drift at discrete observation", *Brazilian Journal of Probability and Statistics*, vol. 29, no. 4, pp. 778–789, 2015.

[KUA 15b] KUANG N., XIE H., "Maximum likelihood estimator for the sub-fractional Brownian motion approximated by a random walk", *Annals of the Institute of Statistical Mathematics (AISM)*, vol. 67, no. 1, pp. 75–91, 2015.

[KUA 16a] KUANG N., "Maximum likelihood estimators for the sub-mixed fractional Brownian motion at discrete observation", *Advances in Mathematics (China)*, vol. 45, no. 3, pp. 471–479, 2016.

[KUA 16b] KUANG N., LIU B., "Least squares estimator for $alpha$-sub-fractional bridges", *Statistical Papers*, pp. 1–20, 2016.

[KUB 17] KUBILIUS K., MISHURA Y., RALCHENKO K., *Parameter Estimation in Fractional Diffusion Models*, Bocconi & Springer Series, vol. 8, Springer, 2017.

[LAN 72] LANDKOF N.S., *Foundations of Modern Potential Theory*, Die Grundlehren der mathematischen Wissenschaften, vol. 180, Springer-Verlag, 1972.

[LAN 05] LANCASTER P., RODMAN L., "Canonical forms for Hermitian matrix pairs under strict equivalence and congruence", *SIAM Review*, vol. 47, no. 3, pp. 407–443, 2005.

[LI 01] LI W.V., SHAO Q.-M., "Gaussian processes: inequalities, small ball probabilities and applications", in *Stochastic Processes: Theory and Methods*, Handbook of Statistics, vol. 19, Amsterdam, 2001.

[LIF 12] LIFSHITS M., *Lectures on Gaussian Processes*, SpringerBriefs in Mathematics, Springer, Heidelberg, 2012.

[LIU 12a] LIU J., PENG Z., TANG D. *et al.*, "On the self-intersection local time of subfractional Brownian motion", *Abstract and Applied Analysis*, Article ID 414195, 2012.

[LIU 12b] LIU J., YAN L., "Remarks on asymptotic behavior of weighted quadratic variation of subfractional Brownian motion", *Journal of the Korean Statistical Society*, vol. 41, no. 2, pp. 177–187, 2012.

[LIU 12c] LIU J., YAN L., PENG Z. *et al.*, "Remarks on confidence intervals for self-similarity parameter of a subfractional Brownian motion", *Abstract and Applied Analysis*, Article ID 804942, 2012.

[LIU 17] LIU J., TANG D., CANG Y., "Variations and estimators for self-similarity parameter of sub-fractional Brownian motion via Malliavin calculus", *Communications in Statistics – Theory and Methods*, vol. 46, no. 7, pp. 3276–3289, 2017.

[MAN 68] MANDELBROT B.B., VAN NESS J.W., "Fractional Brownian motions, fractional noises and applications", *SIAM Review*, vol. 10, pp. 422–437, 1968.

[MAN 97] MANDELBROT B.B., *Fractals and scaling in finance*, Selected Works of Benoit B. Mandelbrot, Springer-Verlag, New York, 1997.

[MAR 06] MARCUS M.B., ROSEN J., *Markov Processes, Gaussian Processes, and Local Times*, Cambridge Studies in Advanced Mathematics, vol. 100, Cambridge University Press, Cambridge, 2006.

[MEL 15] MELNIKOV A., MISHURA Y., SHEVCHENKO G., "Stochastic viability and comparison theorems for mixed stochastic differential equations", *Methodology and Computing in Applied Probability*, vol. 17, no. 1, pp. 169–188, 2015.

[MEN 10] MENDY I., "On the local time of sub-fractional Brownian motion", *Annales Mathématiques Blaise Pascal*, vol. 17, no. 2, pp. 357–374, 2010.

[MEN 13] MENDY I., "Parametric estimation for sub-fractional Ornstein-Uhlenbeck process", *Journal of Statistical Planning and Inference*, vol. 143, no. 4, pp. 663–674, 2013.

[MIA 08] MIAO Y., REN W., REN Z., "On the fractional mixed fractional Brownian motion", *Applied Mathematical Sciences (Ruse)*, vol. 2, nos 33–36, pp. 1729–1738, 2008.

[MIS 08] MISHURA Y.S., *Stochastic Calculus for Fractional Brownian Motion and Related Processes*, Lecture Notes in Mathematics, vol. 1929, Springer-Verlag, Berlin, 2008.

[MIS 11a] MISHURA Y.S., SHEVCHENKO G.M., "Existence and uniqueness of the solution of stochastic differential equation involving Wiener process and fractional Brownian motion with Hurst index $H > 1/2$", *Communications in Statistics – Theory and Methods*, vol. 40, nos 19–20, pp. 3492–3508, 2011.

[MIS 11b] MISHURA Y.S., SHEVCHENKO G.M., "Rate of convergence of Euler approximations of solution to mixed stochastic differential equation involving Brownian motion and fractional Brownian motion", *Random Operators and Stochastic Equations*, vol. 19, no. 4, pp. 387–406, 2011.

[MIS 12] MISHURA Y., SHEVCHENKO G., "Mixed stochastic differential equations with long-range dependence: Existence, uniqueness and convergence of solutions", *Computers & Mathematics with Applications*, vol. 64, no. 10, pp. 3217–3227, 2012.

[MIS 17] MISHURA Y., SHEVCHENKO G., *Theory and Statistical Applications of Stochastic Processes*, ISTE Ltd, London and John Wiley & Sons, New York, 2017.

[MON 07] MONIN A.S., YAGLOM A.M., *Statistical fluid mechanics: mechanics of turbulence. Vol. II*, Dover Publications, Inc., Mineola, 2007.

[NOR 99] NORROS I., VALKEILA E., VIRTAMO J., "An elementary approach to a Girsanov formula and other analytical results on fractional Brownian motions", *Bernoulli*, vol. 5, no. 4, pp. 571–587, 1999.

[NOU 12] NOURDIN I., *Selected Aspects of Fractional Brownian Motion*, Bocconi & Springer Series, vol. 4, Springer, Milan and Bocconi University Press, Milan, 2012.

[NOV 99] NOVIKOV A., VALKEILA E., "On some maximal inequalities for fractional Brownian motions", *Statistics & Probability Letters*, vol. 44, no. 1, pp. 47–54, 1999.

[NUA 02] NUALART D., RAŞCANU A., "Differential equations driven by fractional Brownian motion", *Collectanea Mathematica*, vol. 53, no. 1, pp. 55–81, 2002.

[NUA 06] NUALART D., *The Malliavin Calculus and Related Topics*, 2nd edition, Probability and its Applications, Springer-Verlag, Berlin, 2006.

[OSS 89] OSSIANDER M., WAYMIRE E.C., "Certain positive-definite kernels", *Proceedings of the American Mathematical Society*, vol. 107, no. 2, pp. 487–492, 1989.

[PIT 96] PITERBARG V.I., *Asymptotic Methods in the Theory of Gaussian Processes and Fields*, translated by PITERBARG V.V., Translations of Mathematical Monographs, vol. 148, American Mathematical Society, Providence, 1996.

[PRA 10] PRAKASA RAO B.L.S., *Statistical Inference for Fractional Diffusion Processes*, Wiley Series in Probability and Statistics, John Wiley & Sons, Ltd., Chichester, 2010.

[PRA 12] PRAKASA RAO B.L.S., "Singularity of subfractional Brownian motions with different Hurst indices", *Stochastic Analysis and Applications*, vol. 30, no. 3, pp. 538–542, 2012.

[PRA 17a] PRAKASA RAO B.L.S., "On some maximal and integral inequalities for sub-fractional Brownian motion", *Stochastic Analysis and Applications*, vol. 35, no. 2, pp. 279–287, 2017.

[PRA 17b] PRAKASA RAO B.L.S., "Optimal estimation of a signal perturbed by a sub-fractional Brownian motion", *Stochastic Analysis and Applications*, vol. 35, no. 3, pp. 533–541, 2017.

[REV 91] REVUZ D., YOR M., *Continuous Martingales and Brownian Motion*, Grundlehren der Mathematischen Wissenschaften, vol. 293, Springer-Verlag, Berlin, 1991.

[SAM 93] SAMKO S.G., KILBAS A.A., MARICHEV O.I., *Fractional Integrals and Derivatives*, Gordon and Breach Science Publishers, Yverdon, 1993.

[SAM 94] SAMORODNITSKY G., TAQQU M.S., *Stable Non-Gaussian Random Processes*, Stochastic Modeling, Chapman & Hall, New York, 1994.

[SAM 06] SAMORODNITSKY G., "Long range dependence", *Foundations and Trends in Stochastic Systems*, vol. 1, no. 3, pp. 163–257, 2006.

[SER 88] SERGEICHUK V.V., "Classification problems for systems of forms and linear mappings", *Mathematics of the USSR-Izvestiya*, vol. 31, no. 3, pp. 481–501, 1988.

[SHE 10] SHEN G., YAN L., "The stochastic integral with respect to the sub-fractional Brownian motion with $H > \frac{1}{2}$", *Journal of Mathematical Sciences – Scientific Advances*, vol. 6, no. 2, pp. 219–239, 2010.

[SHE 11] SHEN G., YAN L., "Remarks on an integral functional driven by sub-fractional Brownian motion", *Journal of the Korean Statistical Society*, vol. 40, no. 3, pp. 337–346, 2011.

[SHE 14a] SHEN G., YAN L., "An approximation of subfractional Brownian motion", *Communications in Statistics – Theory and Methods*, vol. 43, no. 9, pp. 1873–1886, 2014.

[SHE 14b] SHEN G., YAN L., "Estimators for the drift of subfractional Brownian motion", *Communications in Statistics – Theory and Methods*, vol. 43, no. 8, pp. 1601–1612, 2014.

[SLE 62] SLEPIAN D., "The one-sided barrier problem for Gaussian noise", *Bell System Technical Journal*, vol. 41, pp. 463–501, 1962.

[STR 84] STRICKER C., "Quelques remarques sur les semimartingales Gaussiennes et le problème de l'innovation", in *Filtering and Control of Random Processes*, Lecture Notes in Control and Information Sciences, vol. 61, Springer, Berlin, 1984.

[TAL 05] TALAGRAND M., *The Generic Chaining*, Springer Monographs in Mathematics, Springer-Verlag, Berlin, 2005.

[TAL 14] TALAGRAND M., *Upper and Lower Bounds for Stochastic Processes*, Ergebnisse der Mathematik und ihrer Grenzgebiete. 3. Folge. A Series of Modern Surveys in Mathematics, vol. 60, Springer, Heidelberg, 2014.

[TES 87] TESTARD F., Polarité, points multiples et géométrie de certains processus gaussiens, PhD thesis, Paris 11, 1987.

[THÄ 09] THÄLE C., "Further remarks on mixed fractional Brownian motion", *Applied Mathematical Sciences*, vol. 38, pp. 1885–1901, 2009.

[THO 91] THOMPSON R.C., "Pencils of complex and real symmetric and skew matrices", *Linear Algebra Application*, vol. 147, pp. 323–371, 1991.

[TUD 07] TUDOR C., "Some properties of the sub-fractional Brownian motion", *Stochastics*, vol. 79, no. 5, pp. 431–448, 2007.

[TUD 09] TUDOR C., "On the Wiener integral with respect to a sub-fractional Brownian motion on an interval", *Journal of Mathematical Analysis and Applications*, vol. 351, no. 1, pp. 456–468, 2009.

[TUD 11] TUDOR C., "Berry-Esséen bounds and almost sure CLT for the quadratic variation of the sub-fractional Brownian motion", *Journal of Mathematical Analysis and Applications*, vol. 375, no. 2, pp. 667–676, 2011.

[TUD 13] TUDOR C.A., *Analysis of Variations for Self-similar Processes*, Probability and its Applications, Springer, Cham, 2013.

[TUD 14] TUDOR C., ZILI M., "Covariance measure and stochastic heat equation with fractional noise", *Fractional Calculus & Applied Analysis*, vol. 17, no. 3, pp. 807–826, 2014.

[VAA 96] VAN DER VAART A.W., WELLNER J.A., *Weak Convergence and Empirical Processes*, Springer Series in Statistics, Springer-Verlag, New York, 1996.

[WAT 44] WATSON G.N., *A Treatise on the Theory of Bessel Functions*, Cambridge University Press, Cambridge, and The Macmillan Company, New York, 1944.

[XIA 08] XIAO Y., "Strong local nondeterminism and sample path properties of Gaussian random fields", in *Asymptotic Theory in Probability and Statistics with Applications*, Advanced Lectures in Mathematics (ALM), vol. 2, Int. Press, Somerville, 2008.

[XIA 09] XIAO Y., "Sample path properties of anisotropic Gaussian random fields", in *A Minicourse on Stochastic Partial Differential Equations*, Lecture Notes in Mathematics, vol. 1962, Springer, Berlin, 2009.

[XIA 11] XIAO W.-L., ZHANG W.-G., ZHANG X.-L., "Maximum-likelihood estimators in the mixed fractional Brownian motion", *Statistics*, vol. 45, no. 1, pp. 73–85, 2011.

[YAN 10] YAN L., SHEN G., "On the collision local time of sub-fractional Brownian motions", *Statistics & Probability Letters*, vol. 80, nos 5–6, pp. 296–308, 2010.

[YAN 11] YAN L., SHEN G., HE K., "Itô's formula for a sub-fractional Brownian motion", *Communications on Stochastic Analysis*, vol. 5, no. 1, pp. 135–159, 2011.

[YEH 14] YEH J., *Real Analysis*, 3rd edition, World Scientific Publishing Co. Pte. Ltd., Hackensack, 2014.

[YIN 15] YIN X., SHEN G., ZHU D., "Intersection local time of subfractional Ornstein-Uhlenbeck processes", *Hacettepe Journal of Mathematics and Statistics*, vol. 44, no. 4, pp. 975–990, 2015.

[ZIL 06] ZILI M., "On the mixed fractional Brownian motion", *Journal of Applied Mathematics and Stochastic Analysis*, Article ID 32435, 2006.

[ZIL 13] ZILI M., "An optimal series expansion of sub-mixed fractional Brownian motion", *Journal of Numerical Mathematics and Stochastics*, vol. 5, no. 1, pp. 93–105, 2013.

[ZIL 14] ZILI M., "Mixed sub-fractional Brownian motion", *Random Operators and Stochastic Equations*, vol. 22, no. 3, pp. 163–178, 2014.

[ZIL 16] ZILI M., "Mixed sub-fractional-white heat equation", *Journal of Numerical Mathematics and Stochastics*, vol. 8, no. 1, pp. 17–35, 2016.

Index

L, M, N

O, P, Q, R

S, T

U, W

Printed in the United States
By Bookmasters